# 3D Printing for Construction in the Transformation of the Building Industry

*Editors*

**Bárbara Rangel**
**Ana Sofia Guimarães**
**João Teixeira**

Faculty of Engineering of University of Porto
Civil Engineering Department
Porto, Portugal

**CRC Press**
Taylor & Francis Group
Boca Raton   London   New York

CRC Press is an imprint of the
Taylor & Francis Group, an **informa** business

A SCIENCE PUBLISHERS BOOK

First edition published 2025
by CRC Press
2385 NW Executive Center Drive, Suite 320, Boca Raton FL 33431

and by CRC Press
4 Park Square, Milton Park, Abingdon, Oxon, OX14 4RN



*Library of Congress Cataloging-in-Publication Data (applied for)*

ISBN: 978-1-032-51771-1 (hbk)
ISBN: 978-1-032-51777-3 (pbk)
ISBN: 978-1-003-40389-0 (ebk)

DOI: 10.1201/9781003403890

Typeset in Palatino Linotype
by Prime Publishing Services

# Preface

As concrete revolutionised construction at the beginning of the XX century, digitalisation is transforming the building industry. The automation of processes is already a reality, and 3D printing is now commonplace not only in industry but also in our homes. Unlike the current industry, Additive Manufacturing (AM) in construction is taking its first steps due to the scale and complexity of the final product. Hence, the research carried out over the last decade points to a revolution in the current construction paradigm, particularly in the organisation of the building process and, therefore, the design methodology since both are strictly connected. The integrated accomplice link between design and manufacturing allows great freedom in form design, whereas various construction phases are dismissed.

Scientific research has explored materials, technologies, product design, and architecture methodologies, helping the industry bring innovation to professional fields. This book provides an overview of state-of-the-art knowledge to understand which path architecture and engineering are following. The latest research achievements are gathered to present the design possibilities that 3D printing for construction (3DC) can offer in the various fields of construction, particularly architecture and engineering.

This book is composed of five main chapters:

- **Chapter 1:** Sustainable Construction: Additive Manufacturing in a Circular Design Framework
- **Chapter 2:** Additive Energy: 3D Printing Thermally Performative Building Elements with Low Carbon Earthen Materials

- **Chapter 3:** From 3D to 5D Printing. Additive Manufacturing of Functional Construction Materials
- **Chapter 4:** 3D-Printed Mortars with Marble Powder Towards Sustainable Construction
- **Chapter 5:** Ceramic AM and Beyond: The Potential of Hybrid Construction Systems

One of the most common words that appears in each work is Sustainability. 3D Printing emerged as a technology in construction and that allows a more sustainable Building Industry.

In Chapter 1 – Sustainable Construction: Additive Manufacturing in a Circular Design Framework – authors explain the way 3DC faces circularity, how it can be used to upcycle waste materials into new classes of building materials and the way it designates new challenges to how circularity is "conceived, performed, and evaluated". This first chapter presents two concepts of material "versatility and material transformation" by which the processing and production of circular 3D construction printing can be assessed. The work revises methods to evaluate circular design in the building industry to "ask how these methods map to the processes of 3D construction printing and how they can be extended to support the assessments of circularity in 3D construction printing". Two case studies are also presented to study circularity through 3D construction printing. It is possible to find how 3DPC is mainly "slurry based and multi-material". New lines for "upcycling waste stream materials in 3DPC suggest how these processes can engage circular ideals. 3D printing with alternative materials remains experimental within architecture and construction, and its technologies are still nascent."

In Chapter 2 – Additive Energy: 3D Printing Thermally Performative Building Elements with Low Carbon Earthen Materials – the authors present several novel design methods and building systems "from the scale of a brick to the scale of a wall" using earth and clay for a combination of simulation-driven design and additive manufacturing. "By leveraging materials readily available in all climates, bespoke, simulation-driven building elements could be manufactured from these low or no-cost materials to create performative, low-carbon buildings. Providing a methodology for material and fabrication-aware energy simulation for additive manufacturing offers a scalable groundwork for future studies across climates and local building requirements."

In Chapter 3 – From 3D to 5D Printing. Additive Manufacturing of Functional Construction Materials – the author says "4D printing (4DP) of functional materials involves exploring new materials and printing techniques to create structures that can perform a range of functions, such as actuation, sensing, and self-assembly. 5D printing (5DP) is an extension of 4DP that adds a fifth dimension—machine learning—to the process, allowing for the optimisation of functional materials to achieve specific desired outcomes. The system can identify the best composition and design for a given application by training algorithms. This chapter presents recent advances in the 4DP and 5DP functional construction materials and discusses the challenges that must be addressed. ML offers opportunities to optimise the printing process and assist in the discovery and optimisation of cement-based materials. Functional construction materials require the addition of special additives (nanoparticles) to the printable mix. ML tools can substantially reduce the experimentation phase, assist in mix development, and provide real-time control, making 4D printing a reality for cement-based materials."

In Chapter 4 – 3D-Printed Mortars with Marble Powder Towards Sustainable Construction – the authors show that using marble powder in concrete is related to the environmental impact of cement production. Adding this waste material to 3D-printed mortars will enhance their environmental performance. It may also improve the printing process and final performance of building components. Several mixtures for 3DP were designed and tested, where Portland lime cement, hydraulic lime and sand were replaced by marble powder. Further characterisation of some mortars in the hardened state (mechanical performance and water absorption) was conducted. Overall, this work underscores the feasibility and benefits of using marble powder in 3D-printed mortars, contributing to sustainable practices, reducing waste, and developing environmentally friendly building materials. As the construction industry continues to embrace sustainability, integrating marble powder in 3D-printed mortars represents a meaningful step towards more eco-conscious and aesthetically diverse construction solutions; however, further optimisation and careful characterisation of the material are still necessary.

Chapter 5 – Ceramic AM and Beyond: The Potential of Hybrid Construction Systems – focuses on utilising digital design and additive manufacturing to create advanced constructive systems based on ceramic materials. The productive system developed promotes the integration of complementary materials to mitigate the weaknesses of the primary material by distributing materials according to the structural scheme achieved through topological optimisation; a hybrid system is created, allowing each material to perform optimally under specific forces.

Design principles for assembly and disassembly are also implemented, enabling this system to be fully reversible and highly repairable. Combining all these materials and processes, from design to production, they can obtain a constructive system that rationalises the type and quantity of materials used, thereby contributing to a more sustainable built environment.

With these contributions, we aim to raise reader awareness of the interest in 3D printing in/for construction, its potential, and its alignment with sustainability concerns.

***Editors***
Bárbara Rangel
Ana Sofia Guimarães
João Teixeira

# Acknowledgements

The authors thank the members of Digi@feup, particularly Manuel Jesus, Sofia Pessoa, Elis Ribeiro and the students of the Master in Industrial and Product Design of the Faculty of Engineering of the University of Porto. And a special thanks to Guilherme Giantini, who helped us at the beginning of this process.

This work was financially supported by: Base Funding – UIDB/00145/2020 of the CEAU – Center for Studies in Architecture and Urbanism, both funded by national funds through the FCT/MCTES (PIDDAC) and the Base Funding – UIDB/04708/2020 and Programmatic Funding – UIDP/04708/2020 of the CONSTRUCT – Instituto de I&D em Estruturas e Construções.

# Contents

Chapter **1**

# Sustainable Construction: Additive Manufacturing in a Circular Design Framework


**Mette Ramsgaard Thomsen, Martin Tamke*,
Gabriella Rossi, Ruxandra-Stefania Chiujdea,
Kate Heywood, Maria Sparre-Petersen
and Paul Nicholas**

CITA Centre for IT and Architecture,
The Royal Danish Academy of Fine Arts, School of Architecture, Denmark

**Figure 1**    Radicant – an investigation of waste stream based 3D printed biopolymer composites. CITA 2022. Photo credit Anders Ingvarsten.

---


*For Correspondence: Martin.Tamke@kglakademi.dk

## INTRODUCTION

The current pressure on the Planetary Boundaries (Rockström et al. 2009) and the acknowledgement of architecture and the impact of the wider built environment will require a step-change in the way we conceive, design and build architecture. For architects and further stakeholders in the built environment to truly participate in overarching climate mitigation strategies, we must find ways of making buildings net zero and nature positive. A central part of this action is to consider the materials with which we build and the processes by which they are produced; the resource they employ; the energy the processes necessitate; and the pollutants they release. Additive manufacture and 3D construction printing has been developing since the mid-1990's. During these thirty years the argument for additive methods as a way of optimising materials and diminishing waste has been part of a larger address for sustainable building. In recent years, this argument has been further extended with a renewed focus on circularity and recycling of building materials. Industry and research is aware that for additive manufacture to retain its relevance and to advance its technologies, it must develop circular principles based on a localisation of material resource, waste stream uptake and closed loop recycling.

This chapter examines how 3D construction printing can incorporate methods of circularity, how it can be used to upcycle waste stream materials into novel classes of building materials and how it brings new challenges to how circularity is conceived, performed and evaluated.

The chapter develops two concepts of material versatility and material transformation as central axioms by which the processing and production of circular 3D construction printing can be assessed. The paper reviews methods of evaluating circular design in the building industry to ask how these methods map to the processes of 3D construction printing and asks how they can be extended to support the assessments of circularity in 3D construction printing.

The chapter presents two explorative case studies developed within CITA (Centre for IT and Architecture) examining circularity through 3D construction printing. The case studies examine 3D printing with upcycled waste stream-based glass (Figure 1) and lignocellulosic reinforced biopolymers (Figure 2) as experimental methods of fabricating architectural elements from waste stream materials. The two case studies are strategically placed on either side of the circular design butterfly diagram, examining both finite materials excavated from the geosphere (glass and silicate) and renewable materials grown within the biosphere (collagen and cellulose). This makes it possible to compare across these resourcing paradigms and understand how material flows impact fabrication, its material composition, processing and final performance.

**Figure 2**   Silica – an exploration of recycled glass.
CITA 2019. Photo credit Maria Sparre-Petersen.

## The Drive towards Circular Design

Circular Design is a transformative approach that aims to fundamentally reshape the way we think about and manage resources. Instead of adhering to the conventional 'take-make-waste' linear economic model, Circular Design emphasizes a paradigm where natural capital is both preserved and enhanced, resource yields are maximized and system risks are minimized by astutely managing both finite stocks and renewable flows (Stahel 2016). Circular Design is based on three principles. It starts from the premise that waste and pollution can be "designed out", right from the conception phase. Secondly, it aims to keep, materials and products in use by emphasizing the importance of extending their utility. This second principle develops methods of repair, refitting and refurbishing products, so as to ensure they remain in circulation for as long as possible, thereby reducing the need for more raw materials. Lastly, Circular Design promotes the restoration and rejuvenation of natural ecosystems to rebuild overall planetary system health (McArthur 2013).

For the construction industry, these principles have profound implications. As the global demand for infrastructure and buildings grows, so does the strain on our natural resources. Circular Design, therefore, offers a path for the industry to become more sustainable and resilient by reconceptualising existing building components and construction materials as valuable resources. Critiques of Circular Design have identified that the focus on components and products, rather than the materials themselves causes a siloing of production and waste (Corvellec et al. 2022). This siloing is evidenced in the Ellen McArthur Foundation insistence that recycling itself is not part of the circular design paradigm (Don't waste group 2023). Recycling and upcycling waste streams are only the extremities of this cascading logics (McArthur 2013).

Additive manufacturing differentiates itself from traditional industrial manufacturing in the way that it is not defined in terms of components and products, but rather has a more interconnected relationship between material and product. The additive manufacturing process supports multiple closed and open loops of material re-use, and can extend circular design through the localisation of material supply chains (Despeisse et al. 2017) and by enabling *in-situ* and distributed recycling (Ford and Despeisse 2016). In the tightest closed loop, sintering-based approaches such as polymer and metal printing, reuse excess powder immediately in subsequent printing cycles. In a further closed loop, additive manufacturing can be used as a recycling tool for waste material such as offcuts and unused polymers that can be recycled into filament for future 3D printing. Significantly, additive manufacturing can also become a means for open loop recycling of parts and materials that were not originally manufactured using additive manufacturing. For example (de Mattos Nascimento et al. 2022) describes turning metal scraps into 3D printing powder for sustainable automotive components. In (Zander et al. 2019), polypropylene sourced from multiple local material sources is blended to make novel 3D printing materials.

In manufacture, closed loop recycling of mono-material is ideal. In contrast, in construction, 3DPC is typically multi-material and slurry based. This opens opportunities beyond closed loop such as upcycling of construction waste and other waste stream into new 3D printed components to address inevitable dissipative losses.

## Slurries in Circular 3D Printed Construction: Concrete, Silicates and Biopolymers

Slurries, in difference to pellets and filaments, are the current paradigm for 3D printing in the construction industry. They allow large scale and large volume printing. We identify three central material systems currently being investigated: concrete, silicates and biomaterials, which leverage the material versatility and transformative properties of slurries to best adhere to a circular design framework within 3D printing for construction.

### Concrete

The most mature method of 3D printed construction (3DPC) is concrete printing. In industry, we find off-the-shelf solutions for large scale

3D printing in the form of industrialised gantry systems for on-site construction as well as prefab set ups (Ma et al. 2022). Here, the guiding argument for 3DPC lies with the additive logic of fabrication allowing for the optimisation of material deployment to "print only what is needed" by enabling higher complexity in the topologies of the building (Khoshnevis 2004).

Emerging research seeks to evaluate the sustainability of these systems by benchmarking them against conventional methods of concrete manufacture (Tinoco et al. 2022). These studies find that across International Life Cycle Data (ILCD) categories *Climate Change, Ecosystem Quality, Human Health and Resources*, the most important contributor to environmental impact is concrete material itself (Roux et al. 2023). This is because of the high content of cement and polluting admixtures needed to control the rheology and curing of the concrete slurry. Additionally, the absence of the aggregate fractions larger than 2.0 mm contribute to increased cement ratio in the slurry (Dey et al. 2022). This leads to their overall increased climate impact.

This problem has led to the exploration of tuning the slurries as such, instrumentalising the material versatility of the recipe logic, to integrate waste stream materials and thereby reduce climate impact. These studies develop strategies for substituting cement with secondary raw materials like burnt shale ashes, plastic waste granules and grinded foam (Butkutė and Vaitkevičius 2023) and introduce alternative fillers such as desert sand, ceramist particles or recycled concrete aggregate (Bai et al. 2021). These efforts follow concepts established for circular building concrete while seeking to carefully adjust the concrete slurry to the 3D print fabrication criteria controlling its rheology, curing and interlayer adhesion as well as its final performance.

## Silicates

Current research examines alternative material classes for 3DPC as viable routes to making 3DPC technologies more sustainable and more circular by deploying less carbon-intensive materials. Here, silicates such as clay and earth based 3DPC have been tested at building scale. By creating clay- and earth-based slurries and tuning their recipe through precise control of humidity, addition of geopolymers and reinforcement fibres, it is possible to print large architectural structures. These studies focus on on-site monolithic as well as component printing with local materials, often with 0-km material provenance. The works of Rael San Fratello (Rael and San Fratello 2019), IaaC (IaaC 2022) and Wasp and Mario Cucinella (Moretti 2023) showcase the potential of 3DPC with earthen materials.

Further silicates to be examined for 3DPC are glass-based. Two general strategies for 3D glass printing emerge: either the glass material is molten and directly extruded into a hot chamber where it cools in a controlled way or the glass material is mixed with a binder, which provides stability during printing at room temperature; and the print is later sintered, which purifies the material into a mono-material glass. The hot extrusion process was pioneered by MIT media lab, which has developed a melting kiln that extrudes clear glass in layers (Klein et al. 2015). A similar layered glass process is currently in the process of commercialisation by the Australian company Maple Glass Printing. Here virgin or recycled glass is processed into several millimetres thick glass filaments, which are then extruded (Maple Glass Printing 2023). This type of process is challenging, due to the need for a high temperature (ca. 1000°C) extrusion head and a high temperature chamber, where the hot glass is cooled down in a controlled way. At Karlsruhe Institute of Technology transparent glass is printed with stereolithography 3D printers at resolutions of a few tens of micrometres. The process uses a photocurable silica nanocomposite that is 3D printed and converted to high-quality fused silica glass via heat treatment (Kotz et al. 2017). In CITA's work, and the first case study of this paper, we examine the upcycling of waste stream based glass dust into 3D printed silicate architectural components using 3D extrusion of a slurry of glass and flammable biological matter (Ramsgaard Thomsen 2020).

So far 3DPC in glass is focused on virgin (Baudet et al. 2019) and high-quality crystal glass and not on the globally most common type: Soda Lime glass, which is typically used for packing of consumer goods and as flat glass for building. While crystal glass types excel in regard to optical properties as they are sought after in many industrial applications, their opportunities for closed-loop recycling are limited. The molecular composition of crystal glass has different coefficient of expansion than other glass types, and different crystal glass types can also differ in coefficient of expansion. This means that crystal glass cannot be recovered in a common recycling stream and should be recovered by the same manufacturing facility that it is produced in.

## Biomaterials

A third alternative material system introduces biomaterials to 3DPC through innovative biopolymer slurries. Here, biopolymer binders such as Chitosan, Xanthan gum, alginates or collagen are combined with bio-based fibres and fillers composing recipes with particular properties that allow 3D printing and steer material performance. Current research deploys different binders with and without fibre reinforcement. What the slurries have in common is that they include high water content that acts as a main solvent and necessitates post 3D print drying. Research at MIT and SUTD has explored Chitosan with or without nano cellulose reinforcement (Dristas et al. 2020, Mogas-Soldevila et al. 2015), CITA has investigated Xanthan gum with paper-based lignocellulose reinforcement (Rossi et al. 2022) as well as collagen glue reinforced with diverse lignocellulose sources including bark, hemp, sugar cane, cotton and seagrass (Nicholas et al. 2023b). University of Minho has explored starch-based recipes reinforced with cellulose (Campos et al. 2021).

The breadth of these studies demonstrates the inherent material versatility of the recipe logic. The ability to change between binders and to include a host of different fibres and fillers from lab grade to waste stream allows these material systems to integrate with circular design principles. In CITA's work, and in the case studies below, we examine the interchangeability of lignocellulose reinforcement strategies with upcycled waste stream-based materials to ideate a new material class with minimal climate impact. The ingredient materials can be sourced locally, reducing their carbon footprint with respect to transportation. The inherent material versatility and interchangeability allows us to respond to local conditions of resource scarcity and abundance. Finally, the printed components remain hygroscopic and receptive to water that can be reintroduced to the material system.

Where these alternative material systems are still nascent, they promise new avenues for how to think, design and produce architecture that is fundamentally related to its context. By using local materials and producing locally, they allow a new sense of resource awareness to both its designer and its user. Furthermore, from an "end of life" perspective, biopolymer both silicates and biopolymer have properties that make them open to being re-fabricated. Their high-water content and ability for water to be reintroduced to the system, makes their curing reversible allowing them to be ground down and serve as resource for their next life or return to biomass through biodegradation.

## Two Cases for a Circular 3D Printed Architecture

The two case studies presented in this chapter examine two classes of these alternative materials for 3DPC. Both studies develop ways of mixing waste stream materials into novel slurries with appropriate viscosity for fabrication and material composition for post fabrication and final material performance. The two cases expand upon the two central concepts developed in this chapter. The glass printing case, silicate, explores the question of material versatility and purity. Here the slurry is transformed during post-processing burning out the binder and creating a de facto mono-material product. At the same time the material is transformed during post-processing changing its molecular structure and becoming silicate. This material transformation ruptures the inherent model of material continuity present in most circular design thinking. In the biopolymer printing case, the focus is on the potential of malleability and changeability that the underlying recipe logic achieves. Here, the differentiation of cellulose fibre reinforcements allows for multi-material printing and steering of performance and aesthetics. While the material remains composite, it embeds a material tolerance that allows for both the uptake of changing waste streams dependent on local availability but also its closed-loop recycled into new products.

The two cases are part of a larger CITA portfolio into 3D construction printing. These works explore both the design logics of 3D printing and their interfacing with architectural design as well as their material make up. Early projects, such as Filigree Robotics (Figure 3a) and Conformal PET Facades (Figure 3b) developed conformal printing methods. In Filigree Robotics these printing processes rethink local crafts processes of over-forming for fine porcelain while in Conformal Facades it is used to locally and strategically reinforce generic panels against wind loading. This interest in combining digital fabrication procedure with crafts-based knowledge has been central to many of CITA's research projects (Ramsgaard Thomsen et al. 2012, Ramsgaard Thomsen and Tamke 2013).

Recent work extends this interest with novel circular material systems. Early works develop geo-to-bio hybrids combining hollow concrete printing with an interior grown mycelium insulation (Figure 3c). The two case studies discussed here explore circular design principles with silvi- and agricultural waste stream materials. Here, most recent studies developed for full scale concrete printing and material grading orchestrating multiple nozzle heads to continuously grade material compositions (Figure 3d). In parallel studies, we are further exploring the translation of these tools to living systems printing with growth medium for bioluminescent bacteria (Figure 3e) and interfacing with post process robotic spraying to coat 3D printed building blocks with a weather protective living skin (Figure 3f).

## Method: Research-by-Design for Design Led Material Probing

The two case studies act as *speculative design probes* that allow ideation for how to deploy a new class of materials. Through design integrated studies, they include material investigation (the composition and mixing of the individual material), the fabrication constraints (the viscosity and control of the material) and the post processing of the material (the firing and drying of the material). They develop specific design languages that steer material deposition in response to the understood fabrication criteria. This results in the evocation of new architectural languages and aesthetics.

**Figure 3** **(a)** Filigree Robotics - conformal printing with porcelain; **(b)** PET Facades - conformal printing with PET on PET sheets; **(c)** Graded 3d printing - experiments with lignocellulose reinforced biopolymer; **(d)** Concrete printing - experiments into concrete-mycellium composites for insulatory performance; **(e)** Imprimer la Lumiere - 3d printing growth medium for bioluminecent behaviour; **(f)** 3d printing Self-Healing Biopolymers - printing with cellulose reinforced biopolymer coacted with bacterial cellulose. Photo credit A, D Anders Ingvartsen, B, C, E, F, CITA.

The case studies follow a research-by-design method (Ramsgaard Thomsen and Tamke 2009) focussed on physical prototyping and experimenting that connect design speculation to technological developments. Design is here understood as a situated engagement that allows the project to connect the material investigation to the contexts of architectural performance and programme while directly questioning how their instrumentation affect our perception of their application (Candy 2019, Schön 2008). The method foregrounds the creation of design-led *material evidence*; embodied physical experiments brought about by engaging technologies of design and fabrication. The project outcomes, the architectural tiles and panels, are therefore not explicitly evaluated for performance but more holistically for the way that enable us to rethink the novel design chains between design refinement, fabrication steering and post-processing.

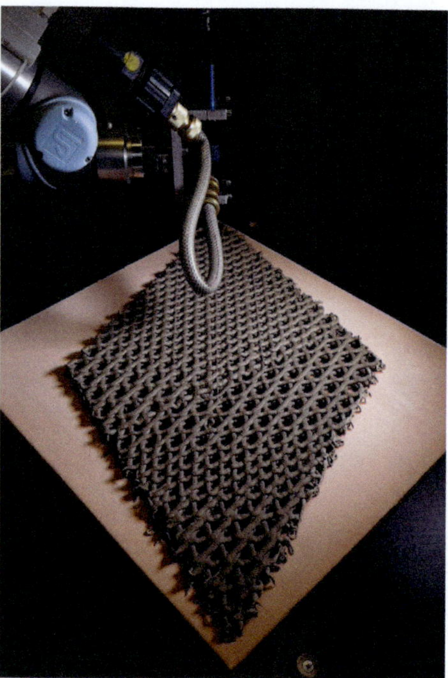

**Figure 4**    Printing of the glass dust based slurry with a robotic arm. Photo credit CITA.

## Case Study 1: Silicate – 3D Printing Recycled Glass

"Silica" is a research probe examining how 3D printing can allow for the upcycling of waste stream-based materials for a circular architecture. By upcycling glass dust, we explore the making of a new type of architectural panels. "Silica" makes use of the ability to print very fine layers thereby enabling a lattice like porous structure. Combining glass crafts with 3D printing, the project explores opportunities, that robotic manufacturing offers for three-dimensional forming otherwise impossible through traditional glass casting techniques.

The project uses the architectural tile as a particular site of investigation, which allows us to contextualise the material investigation within an architectural scenario and explore new architectural expressions. "Silica" is developed as a panelling system by which the tiles are assembled into an architectural skin. Each individual tile is graded in its pattern, thereby locally controlling the porosity and light penetration. When printed in a thin layer our 3D print glass material becomes translucent, when printed in several layers it becomes opaque. This is demonstrated in a screen consisting of 11 individual tiles, where the patterns vary from dense to sparse (Figure 5).

**Figure 5**    The silica tiles with varying patterns steering material density and transparency. Photo credit Maria Sparre-Petersen.

## Slurry Composite and Recipe

The main material in the slurry is glass dust (Figure 6). The glass dust is a waste material derived from the glass recycling industry of soda lime glass typically used in consumer grade glass packaging and flat glass. This glass resource is abundant, cheap and readily available in most parts of the world. Where the two types of Soda lime glass (glass for packaging and flat glass) differ in the application and production method they are both 100% and infinitely recyclable without loss of material qualities. It is a highly recycled material with a recycling rate of 35% in Europe and up to 95% in some countries (Sweden, Belgium and Slovenia) (Furszyfer Del Rio et al. 2022, Glass Packaging Institute 2023).

The glass dust that we are using is a waste stream material resultant from the industrial glass recycling process. As glass is collected for recycling, it is crushed and through this process dust accumulates. In current processing, glass dust cannot be used within this primary recycling process as the airborne dust particles cause problems in the melting furnaces. The glass dust is therefore collected separately and typically downcycled for production of insulation materials or glassphalt neither of which are recyclable.

The glass dust contains impurities from the previous use, such as paper, as well as remnants of the food and beverages, such as yeasts, that need to be considered in the processing and preparation of the material for additive manufacturing. In our recipe the glass dust is mixed with wheat-based baking flour (a starch-based binder) and water into a thick slurry. The water content is tuned to allow for the correct viscosity for extrusion, interlayer adhesion and stability for print stacking.

**Figure 6**    The glass dust waste stream material used in "Silica".
Photo credit Maria Sparre-Petersen.

## Process

The slurry is 3D printed through an extrusion process using a robotic arm. In our process, we have developed bespoke extrusion tools to control over the flow rate. The set-up consists of a robotic arm with an attached tube that feeds the glass slurry (Figure 4). A linear actuator ram feeds the slurry to a ViscoTec dispensing pen attached to the end of the robotic arm. The dispensing pen enables us to stop and start the printing procedure along the print path and gives us control of fine printing.

Once the print is completed, the material is dried and fired in a ceramic kiln. Through experimentation we have found that the best temperature for sintering is 950°C. Where higher degrees improve sintering, the material partially melts and sticks to the separator on the base plate and the 3D print pattern liquifies and becomes indecipherable.

The low sintering temperature causes devitrification. This thermo-chemical process transforms the glass from an amorphous to a crystalline molecular structure (Figure 7). As such, our fired material is no longer classified as glass but as a silicate. However, the material retains its potential for circularity, as the crystalline transformation of soda lime glass is reversible. If fired at high temperatures the material will transform back into glass. Our tiles can therefore be fired and turned back into an amorphous glass state.

The firing process also transforms the material back to a mono-material structure. Where the adding of starch binder to the glass dust makes the multi-material slurry, the firing processes burns out the baking flour binder leaving only the silicate behind. The burning process creates an inherent porosity that can be controlled by varying the recipe of the paste as well as the firing temperature.

The glass waste contains impurities, such as plastic, wood, food and paper. Simple mechanical and thermal processes of sifting and firing of the granulate purify the glass. This provides us with an almost homogenous material base, that does not require an impurity tolerant fabrication process.

**Figure 7**    **(a)** Glass stains allows for printing of coloured elements;. **(b)** Finished print in the kiln. During the drying and firing process, the material contracts as water content evaporates. This process can cause cracking if the material cannot contract uniformly. To prevent this breakage, we have developed 3D print patterns that allow material contraction. Photo credit A, B Simona Hnídková.

Our process provides a means for the small fraction of glass that cannot otherwise be reused with an opportunity to re-enter the closed loop of class recycling instead of being downgraded. The material transformation present in the process is fully reversible allowing the resource to interchange between being silicate and glass. Therefore, the panels remain closed-loop as our 3D prints can at their end of life simply be crushed and printed into new panels or returned to the larger cycle of soda lime glass recycling.

**Figure 8**    CITA print setup multi material printing with lignocellulose reinforced biopolymers. Photo credit CITA.

**Figure 9**    Cellulose Enclosures demonstrator, CITA 2020. Photo credit Anders Ingvartsen.

## Case Study 2: Biopolymer Printing with Waste Stream Materials

The biopolymer case study examines the upcycling of silvi-and agricultural waste stream materials into closed loop 3D printed building components. The case study consists of multiple projects that use different biopolymer binders to produce extruable mixes for 3D printing at large scale (Figures 8, 9 and 10). By mixing the binders with cellulose sources from different waste stream we create novel biopolymer materials that are both fully circular and biodegradable.

The appeal of biopolymer composites lies in the inherent versatility of the recipe logic. The selection of ingredient materials, their proportions as well as the methods of mixing and extruding collectively contribute to the unique properties of the material. Our study investigates the potential for interchangeability with these recipes to steer material performance as well as respond to local material availability.

The projects are based on two main recipe systems. Through the creation of prototypes and demonstrators, we explore their application in indoor architectural scenarios, as self-supporting partitions and as cladding screens. The scenarios create use cases that allow the novel material systems and their fabrication methods to be compared with existing building components.

**Figure 10**  Cellulose screen demonstrator in AEDES exhibition living prototypes. CITA 2022. Photo credit Anders Ingvartsen.

## Slurry Composite and Recipe

Biopolymer composites typically consist of a biopolymer binder that is mixed with a solvent and combined with fibres as reinforcement. As such, the materials are fundamentally multi-material and lock the constituent materials into the binding matrix. In our work we explore two types of binders which are mixed with local waste-stream fibres including bark and wood flour which are often discards from the sawmill industry, cotton fibres which originate from recycled garments, lignocellulose from recycled paper flock insulation, and crushed eelgrass, a typical Danish blue-biomass resource. The first recipe mixes these fibres within an aqueous solution of Xanthan Gum: a polysaccharide binder which is the by-product waste of cheese fermentation and is a common emulsifier and stabilizer used in the food industry (Rech et al. 2022). The second recipe uses a binding matrix made of collagen glue which is a by-product of the treatment of bones and tissues from the meat industry (Figure 11).

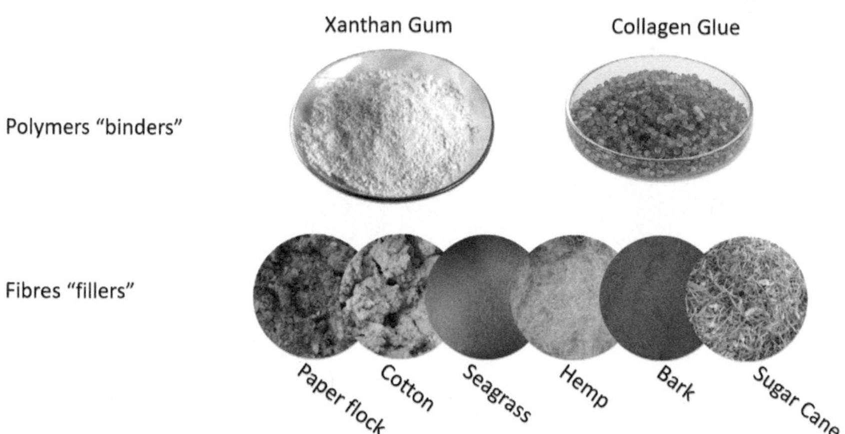


**Figure 11**   Waste stream materials in CITA biopolymer projects.

## Process

The projects deploy custom made extruder systems built for different robotic system setups depending on the size of the target pieces. Moving from the small scale and high versability of the collaborative robots (UR16 robots), across industrial robots (ABB 1600 with a rotary axis) to large scale gantry systems (BOD2), the extruder systems are modified for scale and material composition (Figure 12). Attached to the robotic systems are different extruder and pump types depending on scale and biopolymer binder.

The Xanthan Gum recipe produces a viscous slurry with high structural stability. In our system we interfaced the robotic extrusion with a concrete hopper to control flow rate. The plasticiser in the recipe provides interlayer adhesion during the printing process, which allows us to print up to 40 cm in height at an average speed of 80 mm/s. The recipe is water based, which means that a secondary curing stage occurs after the prints are completed. As the water content of the slurry mixture evaporates, the print hardens and gain strength, but also causes shrinkage of dimensions of the pieces. For that a scalable porous toolpath design strategy using loops has been developed to ensure airflow and evaporation within the printed pieces. This approach, coupled with smart sensing and machine learning workflow, allows for tolerance control and assembly of pieces (Rossi et al. 2021, 2022, 2023).

**Figure 12**   Bespoke 3D print set up across different scales. Photo credit CITA.

The cellulose reinforcement, like the binder, is waste stream based. In the demonstrator project Cellulose Wall and Cellulose Screen we use paper flock as main reinforcement. The paper flock used in the recipe is a recycled paper-based insulation material (Isofloc) a product arriving from the paper recycling industry. Although open loop, paper is highly recycled material source with up to 70% of paper recycled in the US and Europe (CEPI 2022, US EPA 2017) and a recycling rate of up to 8 times (Bureau of International Recycling 2022). The paper floc is highly heterogeneous and coarse with particles of up to 5 mm length. By tuning both the processes of mixing and extrusion, we can accommodate this material coarseness.

The biopolymer material remains fully circular. During printing, wasted slurry can be directly reused in further prints and after drying, the material can be ground down and reused in new slurries. The material is also fully biodegradable and will dissolve in water. This allows the material to return to biomass.

The collagen glue recipe, on the other hand, cannot be extruded cold. Here, collagen pellets are melted into water to form the binding matrix, into which the fibres are dispersed. The material mixture needs to be kept at 60° celsius to maintain extrudable viscosity. This is guaranteed by the usage of a warm pump and heated extruder head. In our projects, the toolpath design allows for the print bead temperature to cool before successive layers are deposited.

In the demonstrator project Radicant (Figure 13), we examine how varying the cellulose reinforcement can grade the material recipe creating different performances. Radicant is a panel based wall screening system. Each panel is created out of several interwoven layers of material creating a highly differentiated multi-material prints. Where these material grading strategies have been explored from a material performance point of view (Rech et al. 2022), the real advantage of the system is the ability to respond to local material availability. In Radicant we use local cellulose-based waste stream materials that are abundant within the Danish industrial ecology. When moving the production to other contexts, the method allows us to replace there cellulose-based waste streams with other materials.

**Figure 13**    The Radicant demonstrator, Aedes Gallery 2022.
Photo credit Anders Ingvartsen.

A central difference between the two recipes is that the curing of the collagen glue based recipe is thermo-reversible. This means that we can reactivate the material forming processes by reintroducing heat and water into the material system. In our work we are examining how this inherent material malleability allows us to engage with the other cascades of Circular Design through practices of repair, refitting and refurbishing (Nicholas et al. 2023a). The thermos-reversibility of the material also allows for direct recycling. By grinding down printed components, we can closed loop recycle the material directly into new slurries. Here, the material grading causes a change to the material composition as otherwise distinct material layers become fused. However, because of their performance compatibility this does not impact the integrity of the prints.

In Radicant, the biopolymer material is therefore also fully circular as the material can be reused in new slurries. The material is also fully biodegradable needing no chemical additives. As with the Xanthan Gum based recipe, this allows the material to return to biomass.

## Towards Evaluation of Circularity in Alternative 3DPC Material Systems

The two cases present new ways by which to understand 3DPC as an upcycling process for architecture. By creating new opportunities of valorising waste stream materials within innovative slurry-based 3D print strategies, they present ways of instrumentalising the inherent material versatility of slurries while suggesting new local and environmentally connected open and closed loop systems that return materials to biomass or to new fully recycled processes. However, the case studies also raise the question of how to evaluate their impact and compare to existing building systems.

Over the last three decades, different frameworks for evaluating circular design in the construction industry have been catalysed and are maturing. These frameworks seek to quantify the environmental impact as well as the circularity of products. While none of these methods have been developed with 3DPC in mind and none of them are directly transferrable for assessing waste-stream based upcycling, some of the guiding concepts and methods could provide ways of evaluating this emerging material and fabrication-based paradigms.

The following overview presents the three central methods for evaluating circularity in architecture and identifies which of the assessments that they enable are relevant for 3DPC and what they lack. Firstly, the Life-Cycle Assessment method, widely used for the climate impact assessment of building components and assemblies. Secondly, the Material Circularity Indicator that evaluates the material flow of building components and assemblies to assess the circularity of their material make-up. And finally, the Recyclability Index that measures the ability for building components and assemblies to be recycled or reused through its decomposition and disassembly.

## Life Cycle Assessment

Life Cycle Assessment (LCA) is the most established environmental sustainability evaluation method used in the built environment. It provides a framework for the assessment of products from raw material extraction to end-of-life processes. LCAs are defined through system boundaries and can be seen as calculating the material flow and fabrication in a linear process going from production (A1-A3), through construction (A4-A5) and use (B1-B7) to end of life (C1-C4) (Hollberg and Ruth 2016). Outside these boundaries is a fifth stage (D) that speculates on recyclability and re-use. While LCA allows for the measurement of multiple axes, the focus within the architecture and construction sector is mainly on the Global Warming Potential (GWP) quantifying the amount of $CO_2$ equivalent emitted during the material's lifespan (Agustí-Juan and Habert 2017).

As 3DPC becomes more prevalent in construction, LCA is being used as the key evaluation method in measuring the impact of these novel material systems and processes. However, LCA for construction has been developed on the premise of standardized, mass-produced building components. This means that its direct application to 3DPC elements is limited (Heywood and Nicholas 2023). 3D printing remains a novel fabrication strategy with few examples of real-world implementation. The inherent versatility of the material recipes and the breadth of the slurries being explored, require new methods by which quantitative data on the printed material can be generated. Current LCA analyses of 3DPC remains experimental, even at production stages, and most of these lack the integration of the fabrication process resulting in incomplete and difficult to compare (Kuzmenko et al. 2020).

LCA or 3DPC with silicates and biopolymers are even further limited. Findings suggest that 3DPC with biopolymers can improve environmental impact by removing the need for energy intensive heating processes and that the slurry materials themselves have lower embodied carbon than comparable printing materials such as ABS (Faludi et al. 2019). There is a general call for further studies for biopolymer LCA that combine print process with embodied material impact including biomass sourcing (Garmulewicz et al. 2023) to be able to create a framework for 3D printing with biopolymers. There is also still a need to quantify the potential circularity of the materials and fabrication process more accurately.

## Material Circularity Indicator

An emerging assessment method for circular design is the Material Circularity Indicator (MCI) created by the Ellen McArthur Foundation

(McArthur n.d.). Based on three principles, the MCI evaluates how restorative the material flow of a product is fraction of input materials sourced from biological sources where the extraction doesn't bypass the restoration capacity; fraction of input materials from recycled or reused sources; potential for extension of the product's lifespan through the reuse, recovery and recycling of materials at the end-of-life and ensuring biological matter remain uncontaminated.

MCI relies on material passport databases that store information on embodied carbon, use and opportunities for circularity (Madaster n.d.). This is a limitation in 3DPC, as databases lack material passport information of innovative slurry mixtures. A second limitation in the MCI method is that it does not consider impact of recycling on the properties of materials. Recycling 3D printed materials is challenging as closed loop processes can lead to material contamination and degradation. This concern is reflected in the concept of 'quality of recycling', defined as the extent to which the distinct characteristics of a material are preserved or recovered so as to maximise their potential to be re-used in the circular economy (European Commission. Joint Research Centre. 2020). Certain material classes are more tolerant of recycling cycles than others. Where instance metals in general have a broad range of use grades, which makes them easier to recycle. In contrast 3D printed polymers can degrade significantly across recycling cycles (Anderson 2017, Kumar et al. 2021).

The MCI methodology does not currently account for upcycling. Waste is considered unrecoverable which is not always the case in 3DPC. Methods for evaluating the valorisation of waste streams, their scarcity and price volatility could support the evaluation of alternative slurry-based material processes.

## Recyclability Index

The Recyclability Index (RI) is used in the construction industry to evaluate the ability of a material to be reused or recycled (Vefago and Avellaneda 2013). Based on statistical entropy principles, RI assesses the material composition of components and evaluates their separability and reuse at end-of-life (Roithner et al. 2022). The RI argues for material systems that are composed of one or few materials where connections are reversible. However, for 3D printing the inherent material versatility of the recipe logic counteracts this metrics. Here, materials are always fused in slurries and the logic of additive fabrication aims to integrate otherwise distinct parts.

However, the experimental work ideas of open loop cycling and biodegradability create new ways of understanding separability and reuse. Rather than seeking to isolate the individual materials, they can be recycled in their entirety by being ground down and remixed into new slurries (Ramsgaard Thomsen et al. 2023)

Where LCA provides a more holistic tool for assessing the environmental impacts of building components and their assemblies, MCI and RI focus more on material flows. By extending MCI to include the valorisation of waste streams the method could more effectively be used for the assessment of a circular 3DPC. In this way MCI could be used as a tool in guiding the selection of filler and binder in early design stages, based on the complementary parameters. That would require an extension of the calculation to include upcycling of waste streams and better-established databases where information on material availability can be accessed. In the same way, RI would need to extend its understanding of material composition and the concept of separability to be implemented for 3DPC.

## Discussion

In Circular Design the concept of entropy is foundational for understanding the limitations of the model. As we build, materials entangle both at the macro scale of composites and the micro scale of atoms creating new compounds that cannot be separated. The guiding principles for circular design therefor focus on separability and reuse. This chapter discusses how 3D printing in construction breaks with these ideals while creating new avenues for thinking circularity in construction.

The chapter has shown how 3DPC is mainly slurry based and multi-material. As such 3DPC locks materials into inseparable composites creating a material complexity that is difficult to undo. However, new avenues for upcycling waste stream materials in 3DPC suggest ways in which these processes can engage circular ideals. The difficulty of evaluating the contribution of upcycling waste streams into new products make it hard to measure the full contribution that 3DPC can have on the way we use resources within the built environment. The case studies show cased in this chapter ask how 3DPC can be a method for innovative rethinking of waste and the creation of new closed and open recycling loops. They demonstrate how the alternative material systems, methods of 3D printing and application can lead to novel architectural solutions.

The case studies also demonstrate how the inherent versatility also leads to a broad material tolerance. As exemplified in the Radicant demonstrator, the ability to replace cellulose reinforcement fibre with new sources allows not only for material grading but also for closed loop cycling. A central application of this versatility is the ability to respond to and leverage local resource availability. This creates an environmental contextualisation of the materials central to sustainability and the Circular Design framework.

3D printing with alternative materials remains experimental within the field of architecture and construction and its technologies are still nascent. However, the inherent agility of the recipe logic and the creative making of bespoke printing sets up at multiple scales allow a new vision for how building materials can be designed, made and deployed.

## ACKNOWLEDGEMENTS

This project has received funding from the European Research Council (ERC) under the European Union's Horizon 2020 research and innovation programme (grant agreement No. 101019693), the Independent Research Fund Denmark, Zukunft Bau and the Danish Arts Foundation. It relies on collaborations with Danish Polymer Centre (Radicant, Cellulose Enclosure, Cellulose Screens) with Anders Egede Daugaard and Arianna Rech.

## REFERENCES

Agustí-Juan, I. and Habert, G. 2017. Environmental design guidelines for digital fabrication. J. Cleaner Prod. 142: 2780–2791. https://doi.org/10.1016/j.jclepro.2016.10.190.

Anderson, I. 2017. Mechanical properties of specimens 3D printed with virgin and recycled polylactic acid. 3D Print. Addit. Manuf. 4(2): 110–115. https://doi.org/10.1089/3dp.2016.0054.

Bai, G., Wang, L., Ma, G., Sanjayan, J. and Bai, M. 2021. 3D printing eco-friendly concrete containing under-utilised and waste solids as aggregates. Cem. Concr. Compos. 120: 104037. https://doi.org/10.1016/j.cemconcomp.2021.104037.

Baudet, E., Ledemi, Y., Larochelle, P., Morency, S. and Messaddeq, Y. 2019. 3D-printing of arsenic sulfide chalcogenide glasses. Opt. Mater. Express. 9(5): 2307–2317. https://doi.org/10.1364/OME.9.002307.

Bureau of International Recycling. 2022. BIR - Paper. Paper Extract Bir Annual Report. 2022. https://www.bir.org/the-industry/paper.

Butkutė, K. and Vaitkevičius, V. 2023. 3D concrete printing with wastes for building applications. Journal of Physics: Conference Series, 2423(1), 012034. https://doi.org/10.1088/1742-6596/2423/1/012034.

Campos, T., Cruz, P.J.S. and Figueiredo, B. 2021. The use of natural materials in additive manufacturing of buildings components. pp. 355–364. *In*: V. Stojakovic and B. Tepavcevic (eds). Towards a New, Configurable Architecture—Proceedings of the 39th eCAADe Conference, Vol. 1. University of Novi Sad.

Candy, L. 2019. Reflection, practive and the creative practitioner. pp. 10–32. *In*: The Creative Reflective Practitioner: Research Through Making and Practice. Routledge & CRC Press. https://www.routledge.com/The-Creative-Reflective-Practitioner-Research-Through-Making-and-Practice/Candy/p/book/9781138632769.

CEPI. 2022. EPRC Monitoring Report 2022. https://www.cepi.org/eprc-monitoring-report-2022/

Corvellec, H., Stowell, A.F. and Johansson, N. 2022. Critiques of the circular economy. J. Ind. Ecol. 26(2): 421–432. https://doi.org/10.1111/jiec.13187.

de Mattos Nascimento, D.L., Mury Nepomuceno, R., Caiado, R.G.G., Maqueira, J.M., Moyano-Fuentes, J. and Garza-Reyes, J.A. 2022. A sustainable circular 3D printing model for recycling metal scrap in the automotive industry. J. Manuf. Technol. Manage. 33(5): 876–892. https://doi.org/10.1108/JMTM-10-2021-0391.

Despeisse, M., Baumers, M., Brown, P., Charnley, F., Ford, S. J., Garmulewicz, A., et al. 2017. Unlocking value for a circular economy through 3D printing: A research agenda. Technol. Forecasting Social Change. 115: 75–84. https://doi.org/10.1016/j.techfore.2016.09.021.

Dey, D., Srinivas, D., Panda, B., Suraneni, P. and Sitharam, T.G. 2022. Use of industrial waste materials for 3D printing of sustainable concrete: A review. J. Cleaner Prod. 340: 130749. https://doi.org/10.1016/j.jclepro.2022.130749.

Don't waste group. 2023. Circular economy vs recycling: What's the difference? Don't Waste Services. https://dontwastegroup.com/waste-management-perspective-circular-economy/

Dristas, S., Vijay, Y., Halim, S., Teo, R., Sanandiya, N. and Fernandez, J.G. 2020. Cellulosic biocomposites for sustainable manufacturing. pp. 74–81. *In*: J. Burry, J. Sabin, B. Sheil and M. Skavara (eds). Fabricate 2020: Making Resilient Architecture. UCL Press. https://doi.org/10.2307/j.ctv13xpsvw.

European Commission. Joint Research Centre. 2020. Quality of recycling: Towards an operational definition. Publications Office. https://data.europa.eu/doi/10.2760/225236.

Faludi, J., Van Sice, C.M., Shi, Y., Bower, J. and Brooks, O.M.K. 2019. Novel materials can radically improve whole-system environmental impacts of additive manufacturing. J. Cleaner Prod. 212: 1580–1590. https://doi.org/10.1016/j.jclepro.2018.12.017.

Ford, S. and Despeisse, M. 2016. Additive manufacturing and sustainability: An exploratory study of the advantages and challenges. J. Cleaner Prod. 137: 1573–1587. https://doi.org/10.1016/j.jclepro.2016.04.150.

Furszyfer Del Rio, D.D., Sovacool, B.K., Foley, A.M., Griffiths, S., Bazilian, M., Kim, J., et al. 2022. Decarbonizing the glass industry: A critical and systematic review of developments, sociotechnical systems and policy options. Renewable Sustainable Energy Rev. 155: 111885. https://doi.org/10.1016/j.rser.2021.111885.

Garmulewicz, A., Tourlomousis, F., Smith, C. and Bolumburu, P. 2023. 3D printing with biopolymers: Toward a circular economy. pp. 371–399. *In*: M. Mehrpouya and H. Vahabi (eds). Additive Manufacturing of Biopolymers. Elsevier. https://doi.org/10.1016/B978-0-323-95151-7.00008-9.

Glass Packaging Institute. 2023. Glass Recycling Facts. https://www.gpi.org/glass-recycling-facts.

Heywood, K. and Nicholas, P. 2023. Sustainability and 3D concrete printing: Identifying a need for a more holistic approach to assessing environmental impacts. Archit. Intell. 2(1): 12. https://doi.org/10.1007/s44223-023-00030-3.

Hollberg, A. and Ruth, J. 2016. LCA in architectural design—A parametric approach. Int. J. Life Cycle Assess. 21(7): 943–960. https://doi.org/10.1007/s11367-016-1065-1.

IaaC. 2022. IAAC builds TOVA: Spain's first 3D printed prototype using earth. IAAC. https://iaac.net/project/3dpa-prototype-2022/

Khoshnevis, B. 2004. Automated construction by contour crafting—Related robotics and information technologies. Autom. Constr. 13(1): 5–19. https://doi.org/10.1016/j.autcon.2003.08.012.

Klein, J., Stern, M., Franchin, G., Kayser, M., Inamura, C., Dave, S., et al. 2015. Additive manufacturing of optically transparent glass. 3D Print. Addit. Manuf. 2(3): 92–105. https://doi.org/10.1089/3dp.2015.0021.

Kotz, F., Arnold, K., Bauer, W., Schild, D., Keller, N., Sachsenheimer, K., et al. 2017. Three-dimensional printing of transparent fused silica glass. Nature. 544: 337–339. https://doi.org/10.1038/nature22061.

Kumar, S., Singh, R., Singh, T. and Batish, A. 2021. On investigation of rheological, mechanical and morphological characteristics of waste polymer-based feedstock filament for 3D printing applications. J. Thermoplast. Compos. Mater. 34(7): 902–928. https://doi.org/10.1177/0892705719856063.

Kuzmenko, K., Gaudillière, N., Feraille, A., Dirrenberger, J. and Baverel, O. 2020. Assessing the environmental viability of 3D concrete printing technology. pp. 517–528. *In*: C. Gengnagel, O. Baverel, J. Burry, M.R. Thomsen and S. Weinzierl (eds). Impact: Design with all Senses. Springer International Publishing. https://doi.org/10.1007/978-3-030-29829-6_40.

Ma, G., Buswell, R., Leal da Silva, W.R., Wang, L., Xu, J. and Jones, S.Z. 2022. Technology readiness: A global snapshot of 3D concrete printing and the frontiers for development. Cem. Concr. Res. 156: 106774. https://doi.org/10.1016/j.cemconres.2022.106774.

Madaster (n.d.) Madaster documentation. Retrieved August 23, 2023, from https://docs.madaster.com/nl/en/knowledge-base/databases

Maple Glass Printing (Director). 2023. 3D Glass Printing | Inside Maple Glass Printing. https://www.youtube.com/watch?v=CSw0Kw6M8i0.

McArthur, E. (n.d.). Material Circularity Indicator (MCI). Retrieved August 23, 2023, from https://ellenmacarthurfoundation.org/material-circularity-indicator.

McArthur, E. 2013. Towards the circular economy, Vol. 1: An economic and business rationale for an accelerated transition. https://www.ellenmacarthurfoundation.org/assets/downloads/publications/Ellen-MacArthur-Foundation-Towards-the-Circular-Economy-vol.1.pdf

Mogas-Soldevila, L., Duro-Royo, J., Lizardo, D., Kayser, M., Sharma, S., Keating, S., et al. 2015. Designing the ocean pavilion: Biomaterial templating of structural, manufacturing, and environmental performance. pp. 1–9. *In*: Proceedings of the International Association for Shell and Spatial Structures (IASS) Symposium 2015, Amsterdam - Future Visions.

Moretti, M. 2023. WASP in the edge of 3D Printing. pp. 57–65. *In*: B. Rangel, A.S. Guimarães, J. Lino and L. Santana (eds). 3D Printing for Construction with Alternative Materials. Springer International Publishing. https://doi.org/10.1007/978-3-031-09319-7_3.

Nicholas, P., Lharchi, A., Tamke, M., Valipour Goudarzi, H., Eppinger, C., Sonne, K., et al. 2023a. Biopolymer composites: Malleable materials for an instable architecture. Acadia23 : Habits of the Anthropocene: Scarcity and Abundance in a Post-Material Economy. pp 166 - 173.

Nicholas, P., Lharchi, A., Tamke, M., Valipour Goudarzi, H., Eppinger, C., Sonne, K., et al. 2023b. Design modeling framework for multi material biopolymer 3d Printing. Proceedings of Advances in Architectural Geometry (AAG) Conference 2023.

Rael, R. and San Fratello, V. 2019. Mud Frontiers. Rael San Fratello. https://www.rael-sanfratello.com/made/mud-frontiers.

Ramsgaard Thomsen, M. and Tamke, M. 2009. Narratives of making: Communicating (by) design 2009. pp. 343–352. *In*: Proceedings of the colloquium 'Communicating (by) Design' at Sint-Lucas Brussels from 15–17th April 2009. Chalmers University of Technology, Sweden.

Ramsgaard Thomsen, M., Tamke, M. and Peder Pedersen, C. 2012. Research 2009–2011—Digital crafting: A network on computation and craft in architecture, engineering and design. pp 2–5. *In*: Research 2009–2011—Digital Crafting (Vol. 1) [Report]. The Royal Danish Academy of Fine Arts, School of Architecture.

Ramsgaard Thomsen, M. and Tamke, M. 2013. Digital crafting: Performative thinking for material design. pp. 242–253. *In*: T. Peters and B. Peters (eds). Inside Smartgeometry: Expanding the Architectural Possibilities of Computational Design. John Wiley & Sons Ltd.

Ramsgaard Thomsen, M. 2020. Silica—A circular material paradigm by 3D printing recycled glass. pp. 613–622. *In*: L. Werner and D. Koering (eds). Anthropologic: Architecture and Fabrication in the Cognitive Age—Proceedings of the 38th ECAADe Conference – Volume 2, TU Berlin, Berlin, Germany, 16–18 September 2020. https://papers.cumincad.org/cgi-bin/works/paper/ecaade2020_128.

Ramsgaard Thomsen, M., Nicholas, P., Rossi, G., Daugaard, A. and Rech, A. 2023. Extending the circular design framework for bio-based materials: Reconsidering cascading and agency through the case of biopolymer composites. pp. 609–620. *In*: M. Ramsgaard Thomsen, C. Ratti and M. Tamke (eds). Design for Rethinking Resources: Proceedings of the UIA World Congress of Ahcitects Copenhagen 2023. Springer Nature Switzerland AG. https://doi.org/10.1007/978-3-031-36554-6_40.

Rech, A., Chiujdea, R., Colmo, C., Rossi, G., Nicholas, P., Tamke, M., et al. 2022. Waste-based biopolymer slurry for 3D printing targeting construction elements. Mater. Today Commun. 33: 104963. https://doi.org/10.1016/j.mtcomm.2022.104963.

Rockström, J., Steffen, W., Noone, K., Persson, Å., Chapin, F.S., Lambin, E.F., et al. 2009. A safe operating space for humanity. Nature. 461: 472–475. https://doi.org/10.1038/461472a.

Roithner, C., Cencic, O., Honic, M. and Rechberger, H. 2022. Recyclability assessment at the building design stage based on statistical entropy: A case study on timber and concrete building. Resour. Conserv. Recycl. 184: 106407. https://doi.org/10.1016/j.resconrec.2022.106407.

Rossi, G., Chiujdea, R., Colmo, C., ElAlami, C., Nicholas, P., and Tamke, M. 2021. A material monitoring framework: Tracking the curing of 3d printed cellulose-based biopolymers. Proceedings of Acadia 2021: Realignements.

Rossi, G., Chiujdea, R., Hochegger, L., Lharchi, A., Nicholas, P., and Tamke, M. 2022. Integrated design strategies for multi-scalar biopolymer robotic 3d printing. Proceedings of Acadia 2022: Hybrids & Haecceities.

Rossi, G., Chiujdea, R.-S., Hochegger, L., Lharchi, A., Harding, J., Nicholas, et al. 2023. Statistically modelling the curing of cellulose-based 3d printed components: Methods for material dataset composition, augmentation and encoding. pp. 487–500. *In*: C. Gengnagel, O. Baverel, G. Betti, M. Popescu, M.R. Thomsen and J. Wurm (eds). Towards Radical Regeneration. Springer International Publishing. https://doi.org/10.1007/978-3-031-13249-0_39.

Roux, C., Kuzmenko, K., Roussel, N., Mesnil, R. and Feraille, A. 2023. Life cycle assessment of a concrete 3D printing process. Int. J. Life Cycle Assess. 28(1): 1–15. https://doi.org/10.1007/s11367-022-02111-3.

Schön, D.A. 2008. The Reflective Practitioner: How Professionals Think in Action. Basic Books, Inc.

Stahel, W.R. 2016. The circular economy. Nature. 531: 435–438. https://doi.org/10.1038/531435a.

Tinoco, M.P., de Mendonça, É.M., Fernandez, L.I.C., Caldas, L.R., Reales, O.A.M. and Toledo Filho, R.D. 2022. Life cycle assessment (LCA) and environmental sustainability of cementitious materials for 3D concrete printing: A systematic literature review. J. Build. Eng. 52: 104456. https://doi.org/10.1016/j.jobe.2022.104456.

US EPA, O. 2017. Paper and Paperboard: Material-Specific Data [Collections and Lists]. https://www.epa.gov/facts-and-figures-about-materials-waste-and-recycling/paper-and-paperboard-material-specific-data.

Vefago, L.H.M. and Avellaneda, J. 2013. Recycling concepts and the index of recyclability for building materials. Resour. Conserv. Recycl. 72: 127–135. https://doi.org/10.1016/j.resconrec.2012.12.015.

Zander, N.E., Gillan, M., Burckhard, Z. and Gardea, F. 2019. Recycled polypropylene blends as novel 3D printing materials. Addit. Manuf. 25: 122–130. https://doi.org/10.1016/j.addma.2018.11.009.

Chapter **2**

# Additive Energy: 3D Printing Thermally Performative Building Elements with Low Carbon Earthen Materials

Alexander Curth*, Eduardo Gascón Alvarez,
Lawrence Sass, Leslie Norford and Caitlin Mueller

MIT Department of Architecture

## INTRODUCTION

Global temperatures continue to rise, driving both rapid urbanization and a resultant widespread need for low-carbon impact, thermally performative, and rapidly scalable building technologies. Solutions combining locally available materials and advances in computational building energy modeling and fabrication offer a potential path toward effective and equitable decarbonization. Additive manufacturing is an emerging technology that enables designers to leverage complex geometry at a low-cost to embed performance across scales. 3D-printed buildings have now been constructed on every continent except Antarctica, and both NASA and SpaceX rely on 3D-printed heat exchange manifolds and

*For Correspondence: curth@mit.edu

functionally graded structural lattices in their rocket engines (Figure 1). To date, few studies have addressed the potential for regulating heat in buildings with additively manufactured elements, in part because of the considerable expense of conventional printing systems and materials. We present a set of novel design methods and building systems from the scale of a brick to the scale of a wall utilizing a combination of simulation-driven design and additive manufacturing with earth and clay. By leveraging materials readily available in all climates, bespoke, simulation-driven building elements could be manufactured from these low or no-cost materials to create performative, low-carbon buildings. By providing a methodology for material and fabrication-aware energy simulation for additive manufacturing, we provide a scalable groundwork for future studies across climates and local building requirements.

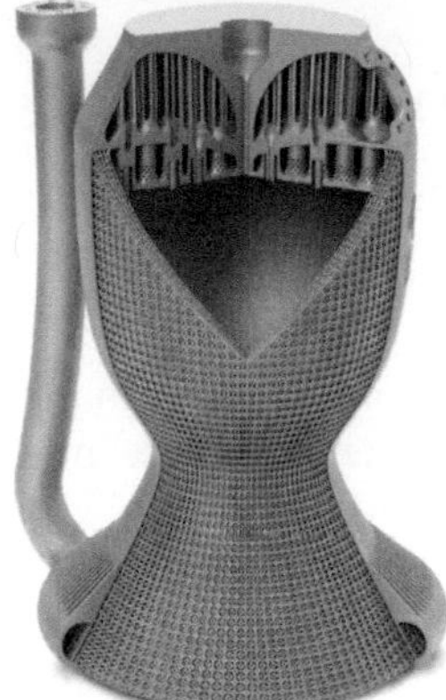 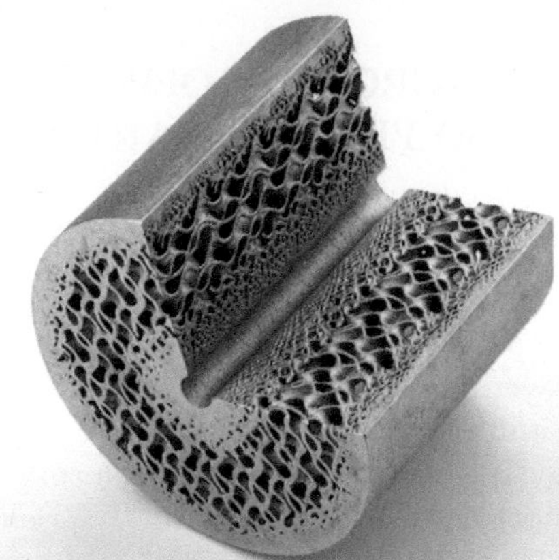

**Figure 1**  High-performance heat exchangers have been 3D printed for aerospace applications including cooling and temperature-regulated fuel injection, demonstrating the potential for highly functional and hierarchical geometric complexity in a monolithic, monomaterial, unit. (Left image – SLM, Right image – Hyperganic).

## IMPACT

As the impact of climate change on cities continues to grow, reducing the cost and time required to produce comfortable and equitable housing is becoming increasingly urgent. Extreme heat events overtax and break urban energy infrastructure from Delhi, India to Houston, USA, resulting in the failure of cooling systems and creating widespread health emergencies. New housing must not only be affordable, it must also include effective means of passive energy regulation, a key design factor in many cultures' vernacular building systems, which has now been replaced with thermally inefficient structures that rely entirely on air conditioning to keep occupants cool. In addition, building under insulated structures without passive thermal comfort performance results in significant embodied energy impact from air conditioning, further exacerbating climate change. At the same time, much of the rapid housing construction happening in cities around the world is concrete, a particularly inefficient and high-carbon material responsible for 8% of global carbon emissions each year (Röck et al. 2020). In this study, we ask if there are broadly scalable methods for additively manufacturing low-carbon, climate-specific architecture that are thermally and structurally performative for housing. We then demonstrate that performative, hierarchical building elements can be produced from extremely low-cost, low-carbon materials like earth using increasingly available 3D printing tools.

## BACKGROUND: OPPORTUNITIES AND CHALLENGES OF 3D PRINTED EARTHEN CONSTRUCTION

### Additive Manufacturing for Construction

The automation of buildings to reduce cost and improve performance has captured the imagination of engineers and inventors since mid-twentieth century. In 1939 William Urschel created the first "Wall Building Machine," a device driven by an analog mechanical cam system to continuously slip-form concrete walls. Over the course of the next decade, Urschel built a series of buildings, including a two-story structure still in use today, eighty years after its construction (Curth 2022). The final iterations of the Wall Building Machine included integrated reinforcement and were used to fabricate structures with complex floorplans and apertures which are still occupied today, overcoming many of the challenges present in the contemporary industry of Large Scale Additive Manufacturing (LSAM).

It would be nearly forty years before the terms "3D printing" or "additive manufacturing" were coined to describe the work of Hideo Kodama (the inventor of selective laser sintering-SLA) and later Chuck Hull (the inventor of powder bed printing) (Kodama 1981, Hull 1986). Despite Kodama's first 3D print being a miniature scale model of a traditional Japanese house, it would not be until the 2000's that additive manufacturing technology would be engaged for architectural scale applications by Berok Khoshnevis (Contour Crafting) and Enrico Dini (D-shape, large scale powder bed printing) (Khoshnevis 2004, Dini et al. 2006). Both Koshnevis's and Dini's approaches have been engaged to make buildings, which, like Urschell's prototypical structures do include some investigation of the integration of mechanical, electrical, and plumbing as well as structural performance and the inclusion of voids to be filled with conventional insulation.

Khoshnevis's approach, Contour Crafting (CC), is the most widespread contemporary method of LSAM. To produce an object, geometry is discretized into layers (contours) which a computer-controlled machine then follows, extruding paste pumped or otherwise supplied to an end effector. Essentially Fused Deposition Modeling (FDM), CC is typically used to make shell structures that can be extruded in a continuous toolpath. To produce more complex geometry with infill, support structures or topologically distinct regions, precise control of extrusion is required to achieve smooth retraction and travel moves. The most advanced systems available today can have the same functionality as a desktop FDM printer, so long as they are supplied with a highly characterized material with consistent properties to facilitate retraction, flow control, and stable mechanical behavior after extrusion. Functionally, this indicated that geometry of similar complexity to the rocket engine components in Figure 1 could be fabricated at an architectural scale if the design tools existed and material properties could be calibrated effectively.

Additive manufacturing allows designers and engineers to create complex geometry quickly and at low cost in a wide range of functional materials. Unlocking this potential for the architectural scale is a growing area of research. Projects in industry and academia have explored myriad materials for architectural AM, including mortar, earth, clay, plastic, and metal. The most common large-scale 3D printed structures are single-story houses constructed from quick-curing mortar. While these structures demonstrate exciting geometric potential, and in some cases, are produced faster than conventional methods would allow, their environmental impact is significantly higher than traditional building systems (Roux et al. 2022, Tinoco et al. 2022). Lower carbon 3D printing systems utilizing earth and clay offer a promising alternative (Curth et al. 2020). The widespread availability of earthen materials suitable for construction makes them a compelling choice for globally scalable AM systems; however, calibrating an un-engineered material like local soil presents unique challenges, including extrusion consistency, shrinkage, and variable mechanical properties. This study presents 3D printed prototypes illustrating thermally performative building elements and 3D printed at high precision with locally sourced earth and clay. Our novel integration of energy modeling and design for additive fabrication allows us to create climate-specific building systems with exceptionally local low-carbon materials. The outputs of our study are designed to be readily adoptable in existing construction practices through adherence to local building energy standards.

## Thermal Performance and Resiliency to Heat

Earthen construction has been used traditionally in temperate and hot desert climates due to its high thermal mass properties, allowing buildings to dampen indoor air temperature fluctuations and avoid excessive hot indoor temperatures (Olgyay 2015). In the current climate crisis context, thermal mass is often revisited as an effective adaptation strategy thanks to the aforementioned peak-reduction effect and its ability to delay in time the consequence of disruptive events such as heat waves or power outages (Zhang et al. 2021). Recent studies have also investigated the ability of earthen assemblies to regulate the moisture content in indoor spaces, a critical aspect when evaluating occupants' thermal comfort conditions (Ben-Alon and Rempel 2023). Moreover, buildings with a high time constant value (i.e., sufficient interior thermal mass and adequate exterior insulation) allow for implementing pre-cooling and pre-heating control strategies that prevent the electric grid from being excessively loaded during high-demand periods, an increasing concern as cities rapidly electrify (Reynders et al. 2013). These multi-domain benefits highlight the importance of finding low-carbon ways to apply thermal mass in buildings as a climate change mitigation and adaptation strategy.

To achieve this, this work investigates how additive manufacturing can further improve the thermal performance of earthen construction (either monolithic walls or discrete blocks) through a context-specific design approach. The fabrication freedom AM provides allows for the precise allocation of the printed material, optimally varying its thermal properties based on the boundary conditions at each surface and the heat flow dynamics within its structure. Yet, recent developments in 3D printing have focused almost exclusively on the elements' structural performance and printability, seldom evaluating their impact on the buildings' energy efficiency and indoor thermal comfort conditions (Pessoa et al. 2021). Existing studies on the thermal performance of 3D printed structures investigate, in most cases, ways to reduce their thermal conductivity through, for example, the design of lattice structures with closed air cavities (Dielemans et al. 2021) or the addition of low-conductivity materials into the mix design (Ma et al. 2022). Other bioclimatic factors have been explored through the construction of prototypes, such as optimal ceramic shading systems (Bechthold et al. 2011) or adaptable heat storage (Sarakinioti et al. 2018). However, very few of these have focused on the abovementioned passive thermal mass performance of earthen buildings (Figliola and Battisti 2021).

There is, therefore, an existing research gap in the design of 3D-printed earthen buildings for optimal and holistic thermal performance while incorporating considerations on material properties, fabrication constraints, and overall environmental impact. As highlighted by a recent review on the heat-moisture properties of 3D printed walls (Li et al. 2023), the combined study of their material and thermal properties opens a promising and relatively unexplored field of research on the multi-functionality aspect of these systems. In this work, we focus on earthen wall systems at a macroscopic scale, exploring printable geometries that maximize the heat resilience benefits earthen materials provide as a low-carbon and accessible resource.

## Research Opportunity

Digital fabrication for architecture has been an active area of academic research for over 20 years; however, the level of adoption by the construction industry is low compared to other manufacturing areas. Digital fabrication is now nearly ubiquitous in the automotive and medical industries while in construction CNC or 3D printed components are only seen in experimental or exceptionally high-end projects like museums or stadiums (Schniederjans 2017). Translating architectural digital fabrication research to the construction site requires acknowledging the tight margins and resulting high risk, associated with adopting emerging technologies for new buildings (Hossain et al. 2020). By using the exceptionally low-cost material of un-engineered soil we aim to create an accessible advance in building technology that allows for the widespread construction of affordable, thermally performative structures. While earth as a building material has critical limitations, strength, and durability when compared to higher-cost materials like concrete or timber, it is available at nearly every building site in the world and, when paired with simulation-driven design, can be shaped into performative geometry across varying climates and programmatic needs. This work aims at testing and expanding these ideas through a novel computational design framework and the prototyping of two earth-based construction systems.

## ADDITIVE ENERGY METHODOLOGY

Additive Energy is a novel design-to-fabrication method that combines digital fabrication, material efficiency, and thermal performance in a holistic and streamlined framework. We use our research site, Santa Barbara, as both the source of our earthen material and the climate around which we designed the prototypical systems demonstrated in the following pages. These include the design of thermally performative wall systems ranging from human assemblable, fired ceramic bricks to raw earth walls 3D-printed on-site at the super-meter scale. We investigated the potential and challenges of both systems for use in residential construction. To do so, we created an optimization loop, linking material and fabrication constraints with design parameters (Figure 2). A thermal analysis then drives the geometry optimization itself to produce an output that balanced material efficiency and thermal performance within the fabrication constraints of our large-scale 3D printing systems.

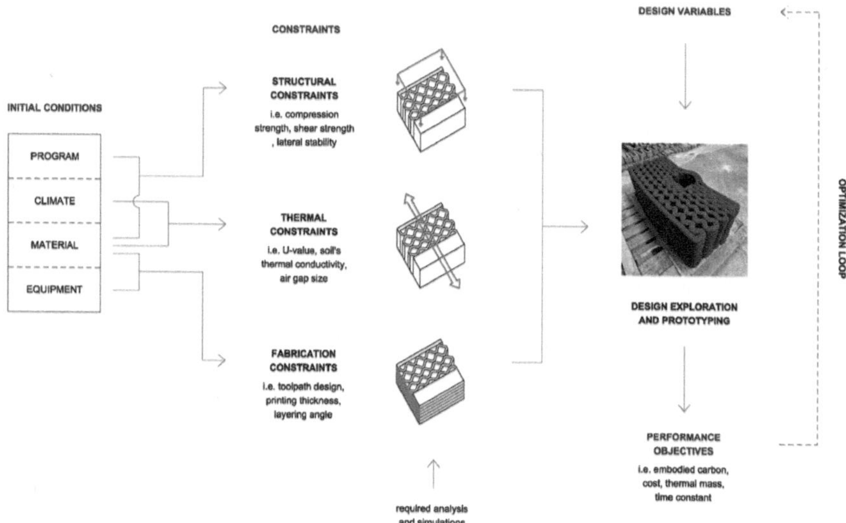


**Figure 2**     Programmatic and climatic conditions are directly incorporated into a material and fabrication-aware digital construction system.

## GEOMETRY DEFINITION AND FABRICATION CONSTRAINTS

Our design optimization methodology balances code-compliant thermal performance with geometries that can be successfully 3D printed at scale with raw earth. To further improve the accessibility of our designs, we constrained them to continuous geometries that do not require expensive in-line valve systems to stop and start extrusion. We use a flexible tool pathing strategy that facilitates both the production of air pockets of varying size throughout a wall section for insulation and solid areas for thermal mass with repeating waveforms of varying frequency and amplitude. Many other geometries exist that could be used to produce a cellular structure for insulation; however, after testing randomly distributed cells, hexagon, circular, and rectilinear void patterns, we settled on a wave-like pattern that facilitates close control of void size and relative position (Figure 3). By controlling the position of adjacent voids, we could adjust position crossing from the exterior to the interior edges of the wall system, tuning its thermal conductivity for performance.

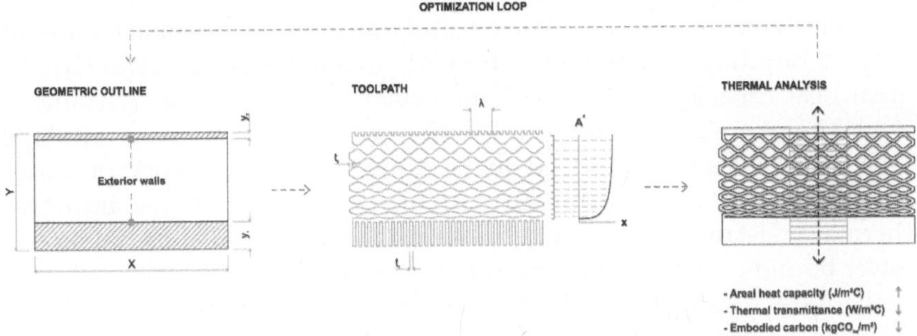


**Figure 3**    A simple wavelike tool pathing strategy facilitates a thermal optimization loop that is inherently fabrication aware.

Figure 3 shows the parametric model used in this work to generate the different wall and block geometries. As observed, given an initial geometric outline, the workflow generates (i) a printable toolpath and (ii) a continuous surface for thermal analysis. While some input parameters are fixed based on fabrication and material constraints (such as wall thickness $t_1$ or total depth $Y$), others become variable parameters input of the mentioned optimization loop (inner wall thickness $y_1$ or the wave amplitude $\lambda$, for example). The metrics $U$ and $\kappa$ (described in the following section) are combined into a single time constant $\tau$ objective and compared against the total embodied carbon in order to understand the trade-offs between thermal performance and material quantities.

## Thermal Analysis

Quantifying the thermal performance of geometrically-complex elements such as the ones presented here is a challenging task, even with existing modern simulation methods. Building Energy Modeling (BEM) engines such as EnergyPlus or TRNSYS typically simplify buildings as prismatic objects with one-dimensional heat flows and, as a result, fail to capture the effects of shaping the elements' surfaces or inner structures. On the other hand, it is possible to characterize the thermal properties of shaped components through numerical methods such as Computer Fluid Dynamics (CFD) or Conjugate Heat Transfer (CHT), yet at the expense of high computational costs and an often slow meshing process. The work presented here alternatively uses analytical models based on first principles and fundamental heat transfer theory as a fast and accurate evaluation tool tailored for early-stage design processes that allows iterating across large numbers of options. While still not fully-validated, early tests show high accuracy between the presented analysis method and their equivalent CHT simulations (2–10% error).

We specifically focus on the analysis of two metrics that allow characterizing of the steady-state and dynamic thermal behavior of shaped building components: thermal transmittance $U$ (W/m²C) and areal heat capacity $\kappa_{in}$ (J/m²C). The former, also defined as $U$-value, is a well-known measure of the insulation properties of a given envelope assembly typically prescribed in building codes. This work proposes computing the $U$-value by discretizing a given geometry into two thermal resistance circuits (one in-series and one parallel as upper and lower bounds, respectively) and calculating a weighted average between both. As a result, the method accounts for the two-dimensional heat transfer processes intrinsic to voided walls and roofs with complex inner structures. The second metric, the areal heat capacity $\kappa_{in}$, expresses the component's thermal mass ability as the periodic heat flow that flows into its internal surface through a 24 h cycle. It is computed by applying the standard ISO 13786 for the dynamic thermal performance of building components (ISO 2007). Both metrics are finally synthesized into a combined time constant value $\tau(\kappa/U)$ that captures their combined benefits: as $\tau$ increases, the component's ability to dampen and time-shift indoor temperature fluctuations increases.

## RESULTS

The following pages present an in-depth study of two shaped building elements with homogenous thermal properties. Two different scales – a discrete brick and a continuous wall – are analyzed with two different earth-based materials – clay and soil – to demonstrate the applicability of the ideas and methods presented in the previous section. The proposed building systems are specifically designed for the climate and available local materials in Santa Barbara, California. This coastal city has a warm-summer Mediterranean climate – zone CSb according to the Köppen-Geiger classification (Beck et al. 2018) – with large diurnal swings and average temperatures that fall within comfort ranges during most of the year. These conditions make thermal mass an ideal passive cooling strategy to explore in the context of low-carbon architecture.

### Additive Block

Multi-cellular clay blocks are an increasingly popular solution in masonry construction thanks to their improved thermal transmittance values, allowing for a reduction (or, in some cases, total substitution) of conventional insulation materials such as mineral wool or extruded polystyrene. This system has been implemented as an envelope solution in contemporary low-energy buildings, more noticeably in the 2226 office building designed by Baumschlager Eberle (Maierhofer et al. 2022). Here we investigate the possibility of further improving the thermal performance of multi-cellular clay blocks at a minimal material cost by leveraging AM technologies and shape optimization methods. This geometric exploration is conducted through a multi-objective optimization (MOO) process based on the parametric model described in Figure 2 for given material properties (medium fire stoneware clay) and fabrication constraints (4 mm printing nozzle and fixed 15 cm brick depth).

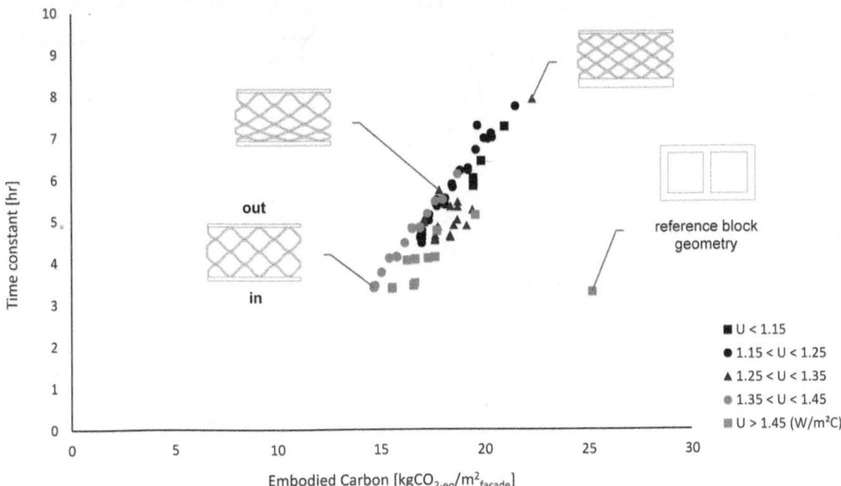


**Figure 4**   The mass customization inherent to the AM process facilitates a gradient of climate-specific designs.

Two objectives are defined: (i) minimizing the embodied carbon per square meter of the facade and (ii) maximizing the described time constant metric $\tau$, which accounts for the heat capacity and the $U$-value simultaneously. Figure 4 illustrates how, as initially expected, both metrics counteract each other: adding more material into the brick's inner structure improves its time constant – increasing its storage capacity and number of air pockets – at the expense of a higher embodied impact. The resulting Pareto front reveals a spectrum of sub-optima solutions that, in all cases, have lower embodied carbon (up to 44%) and higher time constant (up to 139%) than a standard reference brick with two rectangular voids. These results highlight the importance of material distribution and geometry optimization in achieving lightweight solutions with enhanced thermal performance. As observed in Figure 4, the obtained bricks with higher time constant values present a gradient in their void structure, with smaller air pockets on the brick's inner side (increasing its internal heat capacity $\kappa$ where it is more needed) that gradually become larger as they approach the outer side (reducing the needed material).

The fabrication of this optimized brick serves as an opportunity to investigate the inclusion of additional thermal features as a further step toward multi-functionality. In this case, an outer shell is attached to the element's exterior face to provide shade, minimizing the heat gains through solar radiation and providing a ventilation chamber for heat dissipation. A minimum opening area of 50 cm² for every meter of brick and a gap size of 20 mm is ensured, following existing façade design guidelines (Herzog et al. 2012). This ventilation chamber is achieved by tilting the outer shell (in this case, with an angle of 8°) vertically, providing an effective water-proofing layer similar to rain screens in ventilated facades. Figure 5. shows the printing process of one design iteration through a continuous toolpath that includes interior thermal mass, insulation, solar shading, and brick interlocking.

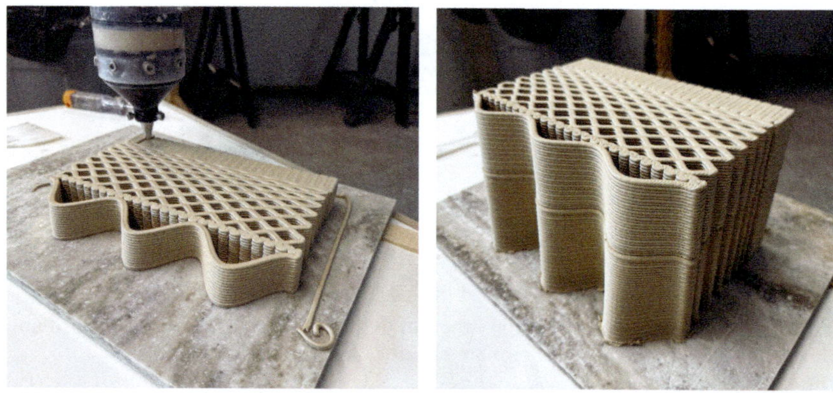

**Figure 5**  A 3D-printed ceramic block contains a multitude of thermal comfort functions in a monolithic unit; solar shading, insulation, and thermal mass. Manufactured to standard extruded block sizes, this prototype can be assembled with common practice masonry methods.

## Additive Wall

The second case study focuses on designing and prototyping multi-functional earthen walls as a low-carbon, local construction method. From a thermal point of view, this research aims at improving the thermal mass and insulation properties of earthen walls through geometric modifications on their internal structure. Unlike traditional monolithic systems, AM allows for fabricating wall components with intricate geometries, resulting in longer heat conduction paths and embedded air pockets that provide high insulation at zero cost. The opportunity then becomes to leverage these strategies to design envelope solutions that meet the energy efficiency standards for the specific context of Santa Barbara. Wall geometries are generated using the previously presented parametric model, accounting for the wall's fabrication constraints and local soil's material properties. More importantly, we define a minimum printing thickness of 3 cm, given the larger nozzle needed to pump the soil mix, and a total wall depth of 50 cm.

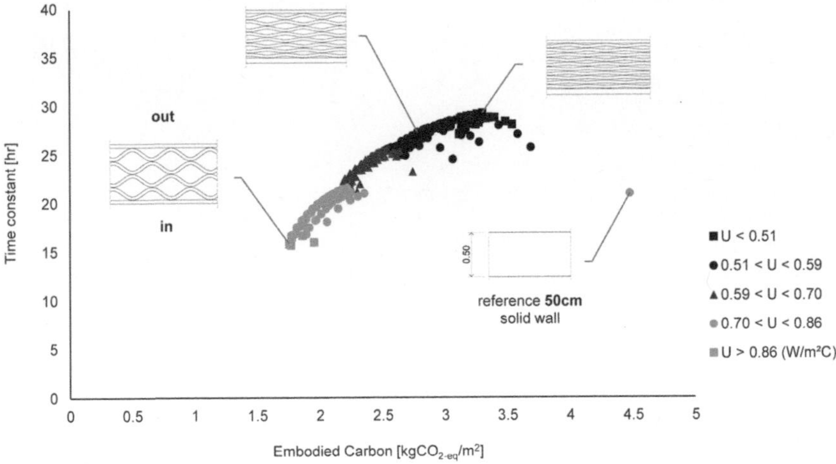


**Figure 6**  The obtained Pareto allows for selecting designs with minimal embodied carbon for a given target time constant value (which defines the wall's ability to dampen indoor temperature fluctuations).

A MOO process is conducted, once again, to maximize the time constant and minimize the embodied carbon. Figure 6 shows the obtained results. As observed, the shaped wall patterns achieve significantly higher time constant values (up to 40%) with reduced embodied carbon emissions (at least 27%) relative to a reference 50 cm solid wall. These results are particularly relevant from a heat resilience perspective, as they unlock the possibility of designing envelope systems that dampen indoor air temperature fluctuations more effectively and provide a slower response time to extreme heat events. At the same time, the *U*-value is improved drastically across all geometries, reaching, in most cases, values that comply with many of the geographic zones defined in the ASHRAE energy standard 90.1 (ASHRAE 2016). For the specific climate of Santa Barbara (zone 3), designs with *U*-values lower than 0.59 W/m2C are sufficient for wall assemblies. Other formulations of the MOO problem could include this condition (*U*-value < 0.59 W/m2C) as a constraint, ensuring that the process generates code-compliant designs exclusively.

A Pareto-optimal design was printed along a curved wall to test the system's ease of fabrication, paying attention to shrinkage issues during the drying process and dimensional stability. These early tests revealed the suitability of including expansion joints when printing long segments of walls. Further, the toolpath included the formwork for casting a concrete column as a main vertical structural element (Figure 7).

This solution proved the suitability of adding structural reinforcements into the wall with a continuous "insulation layer" that avoids any possible thermal bridge.

**Figure 7** Raw earth walls can be 3D printed to climate-specific geometric configurations of thermal mass and insulative voids. Here a 1 × 0.5 × 0.5 wall section demonstrates the functional layout of material for thermal performance in Santa Barbara, California, as well as integrated formwork for a reinforced concrete column.

## DISCUSSION

### Low Carbon Thermal Comfort and Fire Resistance

The enhanced thermal performance of both systems responds to three phenomena: adding air pockets for insulation, longer thermal conduction paths, and the optimized distribution of materials across the wall section for thermal mass purposes. Combining these concepts with a simulation-informed optimization loop allowed us to quickly produce a climate-specific set of solutions at multiple scales, a brick, and a wall section. Both prototypes embody contemporary advances in building energy design while also being exceptionally low carbon and materially efficient. Unlike conventional, discrete building systems of hierarchical elements (rain screen, insulation, vapor barriers, structure), our system embeds necessary functions in a monomaterial fashion, is quick to fabricate, and is generally independent of the complex material supply chains of modern construction. The key takeaway from our study is that a readily accessible, low-cost, low-carbon material can be transformed into a highly performative modern building element through additive manufacturing for a wide range of climates.

In the specific case of Southern California, these prototypes not only respond to a need for passive thermal comfort, but to a growing demand for architecture that is both low embodied carbon and fire resilient. As climate change exacerbates California's annual cycle of wild fires to point of widespread urban impact (Radeloff et al. 2018), new building systems are required to minimize future urban fire risk. Common practice today involves rebuilding burned homes with concrete, a material that adds to the climate impact driving an increased wildfire season. Offering an energy efficient and thoroughly fire resistant alternative could make for greater urban fire resilience.

## Future Work

The key questions resulting from this work are focused around scalability, both in terms of the printing systems used and industry/governmental willingness to adapt to a radical new way of building. Unlike conventional earthen building systems like rammed earth (pisé) or Compressed Earth Blocks (CEB), the prototypes created in this study have the potential to require minimal labor to fabricate at the architectural scale, opening new opportunities for the implementation of earth architecture.

Over the past decade, a number of 3D-printed mortar wall systems have been added to international building codes (ICC Digital Codes 2021). However, it is the inherent advantages of 3D printing for construction that make it difficult to standardize. Different companies use a wide range of proprietary materials, validation software, and mass customizable geometries, making the technology somewhat difficult to fit into existing codes or engineering standards. Our work addresses this challenge through direct and rigorous simulation and testing of a given material.

In addition to challenges of policy and industry perception, many technical as well as Life Cycle Assessment (LCA) aspects of additive energy design remain to be explored. The next steps in the development of both the block and the wall system are laboratory tests of multiple samples to determine the thermal conductivity and time constant of the dried and/or fired prototypes to calibrate the models described to the specific material properties of locally sourced earth and clay. Increasingly complex geometries could also be developed to further reduce the thermal conductivity of what are essentially earthen metamaterials (a single material with varying characteristics). Furthermore, the energy performance of novel 3D printed building systems could also be balanced with other critical metrics, such as mechanical performance. Figure 8 shows an early test by the authors of a closed-cell, ceramic metamaterial designed to insulate while providing a structural building unit. As geometric complexity increases so do the challenges of simulating the behavior of a given system, making the space for further research broad.

**Figure 8** A ceramic or earth cellular solid can be functionally graded and parameterized to balance thermal and structural performance, creating a highly functional building element from one of the world's most readily available low-carbon materials.

## CONCLUSION

In this study we present novel methods combining thermal simulation, optimization, and low carbon 3D printing at an architectural scale. The result is a series of prototypes which demonstrate potential for the creation of wall systems that are low carbon, thermally performative and quick to produce with low cost, locally sourced earthen materials. In addition, the Additive Energy methodology is parameterized to allow for the production of climate specific geometry, balancing insulation and thermal mass as needed. Most critically, these systems are compared to local thermal and structural requirements, offering the first steps in a path to industry adoption and ultimately a reduction of the carbon impact of future construction.

## ACKNOWLEDGMENTS

Thank you to Clay Studio Santa Barbara and LYNDA Labs for providing space to conduct these experiments and the MIT Programmable Mud Initiative for providing materials and equipment.

## REFERENCES

Appendix AW 3D-Printed Building Construction. 2021. International Residential Code (IRC) | ICC Digital Codes. 2021. ICC Digital Codes. https://codes.iccsafe.org/content/IRC2021P2/appendix-aw-3d-printed-building-construction.

ASHRAE, ASHRAE Standard. 2016. Standard 90.1-2016; Energy Standard for Buildings Except Low Rise Residential Buildings. American Society of Heating, Refrigerating and Air-Conditioning Engineers, Inc.: Atlanta, GA, USA.

Bechthold, M., King, J., Kane, A., Niemasz, J. and Reinhart, C. 2011. Integrated environmental design and robotic fabrication workflow for ceramic shading systems. pp. 70–75. *In*: Kwon, S. (ed.). Proceedings of the 28th ISARC, Seoul, Korea. International Association for Automation and Robotics in Construction (IAARC). https://doi.org/10.22260/ISARC2011/0010.

Beck, H.E., Zimmermann, N.E., McVicar, T.R., Vergopolan, N., Berg, A. and Wood, E.F. 2018. Present and future Köppen-Geiger climate classification maps at 1-km resolution. Scientific Data 5(1): 180214. https://doi.org/10.1038/sdata.2018.214.

Ben-Alon, L. and Rempel, A.R. 2023. Thermal comfort and passive survivability in earthen buildings. Build. Environ. 238: 110339. https://doi.org/10.1016/j.buildenv.2023.110339.

Curth, A., Darweesh, B., Arja, L. and Rael, R. 2020. Advances in 3D printed earth architecture: On-site prototyping with local materials. pp. 105–110. *In*: BE-AM: Building Environment Additive Manufacturing. Damstadt.

Curth, A. 2022. Slip forming to construction 3D printing: the early history of large-scale additive manufacturing. ICI Journal. 23: 16–20.

Dielemans, G., Briels, D., Jaugstetter, F., Henke, K. and Dörfler, K. 2021. Additive manufacturing of thermally enhanced lightweight concrete wall elements with closed cellular structures. J. Facade Des. Eng. 9(1): 59–72. https://doi.org/10.7480/JFDE.2021.1.5418.

Dini, E., Nannini, R. and Chiarugi, M. 2006. Method and device for building automatically conglomerate structures. World Intellectual Property Organization WO2006100556A2, filed March 16, 2006, and issued September 28, 2006. https://patents.google.com/patent/WO2006100556A2/en.

Figliola, A. and Battisti, A. 2021. Informed architecture and clay materials. Overview of the main european research paths. pp. 47–72. *In*: Figliola, A. and Battisti, A. (eds). Post-industrial Robotics. Springer, Singapore. https://doi.org/10.1007/978-981-15-5278-6_2.

Herzog, T., Krippner, R. and Lang, W. 2012. Facade Construction Manual. Walter de Gruyter.

Hossain, Md. Aslam, Zhumabekova, A., Paul, S.C. and Kim, J.R. 2020. A review of 3D printing in construction and its impact on the labor market. Sustainability 12(20): 8492. https://doi.org/10.3390/su12208492.

Hull, C.W. 1986. Apparatus for production of three-dimensional objects by stereolithography. United States US4575330A, filed August 8, 1984, and issued March 11, 1986. https://patents.google.com/patent/US4575330A/en.

ISO, EN. 2007. 13786: 2007—Thermal Performance of Building Components—Dynamic Thermal Characteristics—Calculation Methods. CEN. European Committee for Standardization 27.

Khoshnevis, B. 2004. Automated construction by contour crafting—related robotics and information technologies. Automation in Construction, The best of ISARC 2002, 13(1): 5–19. https://doi.org/10.1016/j.autcon.2003.08.012.

Kodama, H. 1981. Automatic method for fabricating a three-dimensional plastic model with photo-hardening polymer. Rev. Sci. Instrum. 52(11): 1770–1773. https://doi.org/10.1063/1.1136492.

Li, Z., Xing, W., Sun, J. and Feng, X. 2023. Multiscale structural characteristics and heat–moisture properties of 3D printed building walls: A review. Constr. Build. Mater. 365: 130102. https://doi.org/10.1016/j.conbuildmat.2022.130102.

Ma, G., Ruhan A, Xie, P., Pan, Z., Wang, L. and Hower, J.C. 2022. 3D-printable aerogel-incorporated concrete: Anisotropy influence on physical, mechanical, and thermal insulation properties. Constr. Build. Mater. 323: 126551. https://doi.org/10.1016/j.conbuildmat.2022.126551.

Maierhofer, D., Röck, M., Saade, M.R.M., Hoxha, E. and Passer, A. 2022. Critical life cycle assessment of the innovative passive nZEB building concept 'be 2226' in view of net-zero carbon targets. Build. Environ. 223: 109476. https://doi.org/10.1016/j.buildenv.2022.109476.

Olgyay, V. 2015. Design with Climate: Bioclimatic Approach to Architectural Regionalism, New and expanded edition. Princeton: Princeton University Press. https://doi.org/10.1515/9781400873685.

Pessoa, S., Guimarães, A.S., Lucas, S.S. and Simões, N. 2021. 3D printing in the construction industry - A systematic review of the thermal performance in buildings. Renewable Sustainable Energy Rev. 141: 110794. https://doi.org/10.1016/j.rser.2021.110794.

Radeloff, V.C., Helmers, D.P., Kramer, H.A., Mockrin, M.H., Alexandre, P.M., Bar-Massada, Avi., et al. 2018. Rapid growth of the US wildland-urban interface raises wildfire risk. PNAS. 115(13): 3314–3319. https://doi.org/10.1073/pnas.1718850115.

Reynders, G., Nuytten, T. and Saelens, D. 2013. Potential of structural thermal mass for demand-side management in dwellings. Build. Environ. 64: 187–199. https://doi.org/10.1016/j.buildenv.2013.03.010.

Röck, M., Saade, M.R.M., Balouktsi, M., Rasmussen, F.N., Birgisdottir, H., Frischknecht, R., et al. 2020. Embodied GHG emissions of buildings – The hidden challenge for effective climate change mitigation. Appl. Energy. 258: 114107. https://doi.org/10.1016/j.apenergy.2019.114107.

Roux, C., Kuzmenko, K., Roussel, N., Mesnil, R. and Feraille, A. 2022. Life cycle assessment of a concrete 3D printing process. Int. J. Life Cycle Assess. 28: 1–15. https://doi.org/10.1007/s11367-022-02111-3.

Sarakinioti, M.-V., Konstantinou, T., Turrin, M., Tenpierik, M., Loonen, R., De Klijn-Chevalerias, M.L., et al. 2018. Development and prototyping of an integrated 3D-printed façade for thermal regulation in complex geometries. J. Facade Des. Eng. 6(2): 029–040. https://doi.org/10.7480/JFDE.2018.2.2081.

Schniederjans, D.G. 2017. Adoption of 3D-printing technologies in manufacturing: A survey analysis. Int. J. Prod. Econ. 183: 287–298. https://doi.org/10.1016/j.ijpe.2016.11.008.

Tinoco, M.P., de Mendonça, É.M., Fernandez, L.I.C., Caldas, L.R., Reales, O.A.M. and Filho, R.D.T. 2022. Life cycle assessment (LCA) and environmental sustainability of cementitious materials for 3D concrete printing: A systematic literature review. J. Build. Eng. 52: 104456. https://doi.org/10.1016/j.jobe.2022.104456.

Zhang, C., Kazanci, O.B., Levinson, R., Heiselberg, P., Olesen, B.W., Chiesa, G., et al. 2021. Resilient cooling strategies – A critical review and qualitative assessment. Energy Build. 251: 111312. https://doi.org/10.1016/j.enbuild.2021.111312.

Chapter **3**

# From 3D to 5D Printing: Additive Manufacturing of Functional Construction Materials

**Sandra S. Lucas**
https://orcid.org/ 0000-0003-3893-5322

Eindhoven University of Technology,
Department of the Built Environment, Eindhoven, the Netherlands
23_TF_3DPCAEC_lucas_ch_DPC.docx

## INTRODUCTION

Additive manufacturing has revolutionised the industry by allowing the production of more complex geometries and easier customization of objects. One area where 3D printing is proving to be of immense potential is in the construction sector (Ahmed 2023, Guimarães et al. 2021). Traditional construction often involves labour-intensive processes and is limited in terms of design freedom and customization. For the building industry, 3D printing means the possibility of manufacturing intricate structures and components with reduced construction time and costs. Researchers and engineers have been looking into ways to

---

*For Correspondence: s.s.d.o.lucas@tue.nl

further improve 3D printing's capabilities in the construction industry because of this promising initial success. One feature that is now being investigated is the introduction of a fourth dimension to the printing process, a concept that introduces aspects of functionality and responsiveness to 3D-printed objects (Huang et al. 2023). In 4D printing, functional cement-based materials are designed to exhibit dynamic properties that allow them to act as sensors and self-clean, or self-repair. (Lucas et al. 2018, Lucas and Barroso de Aguiar 2019). With 4D printing, architects and engineers can create structures that adapt to their environment, withstand external forces, and perform complex functions (Alves et al. 2023).

Construction materials have always been developed using trial-and-error methods. Adding functionalities to materials introduces a new layer of complexity to the development of printable compositions that is incompatible with this traditional 'Edisonian' approach (Xue et al. 2016). To overcome the limitations of experimental methodologies in materials optimisation, researchers traditionally resort to the use of computational modelling. However, every model developed so far fails, at some level, to realistically describe the materials' behaviour (Zhang et al. 2020). Cement undergoes extraordinarily complex processes, including multiple chemical reactions at different time-length scales (some still not fully understood or explained). The complexity and inherent variability of cement materials make it challenging for models to capture all the nuances of the hydration process; therefore, these simulations can potentially render unfit and misleading results (Heinz and Heinz 2021). The quality and representativeness of the training dataset add another level of uncertainty to these simulations. Therefore, while computational modelling can significantly enhance the efficiency of material discovery for additive manufacturing, experimental validation remains crucial to ensure the reliability and performance of the developed cement compositions.

Machine-learning algorithms have the potential to overcome the limitations of computational modelling in accurately predicting the behaviour of cement materials during the hydration process (Zaki et al. 2022). By training the algorithms on a large dataset of experimental results, we can learn the complex relationships between composition, processing parameters, and material properties. This enables the algorithms to rapidly screen and optimise new cement compositions for additive manufacturing, ultimately speeding up the discovery of functional materials (Kekez and Kubica 2020).

In this chapter, after introducing the reader to the extrusion-based manufacturing of cement mortars and the concept of (multi)functional construction materials (4D printing), we discuss how machine-learning algorithms can be used to accelerate the discovery of new cement-based compositions and optimise the manufacturing process (5D printing) (Milazzo and Libonati 2022).

## OVERVIEW OF 3D PRINTING OF CEMENT-BASED MATERIALS

Even though the construction industry has always been a slow adopter of new technologies, additive manufacturing is attracting interest in the sector, and it is already possible to find some examples of commercial solutions (Chang et al. 2023). While the technology is not yet fully developed, its potential is clear, ranging from design freedom to low dependency on human labour, increased safety, and faster production. Additive manufacturing in the construction industry was first introduced with the main objective of creating complex structures that were not possible using the standard formwork method. With the development of the technology, other benefits became apparent, including the reduced need for human labour, which can lead to cost savings, and increased efficiency in the construction process, which can reduce material use and waste. As technology continues to develop and improve, it is expected to revolutionise the construction industry and pave the way for more innovative and sustainable building solutions.

**Figure 1**   Gantry robot system at the University of Eindhoven, TU/e. (https://www.tue.nl/en/research/research-groups/ structural-engineering-and-design/3d-concrete-printing)

Additive manufacturing covers a broad range of methods; extrusion-based 3D printing is one of the most widely employed ones when printing cement mortars or concrete (Altıparmak et al. 2022). As shown in Figure 1, a robotic arm or gantry system equipped with motors and actuators allows movement in multiple directions (X, Y, and Z axes), accurately positioning the nozzle in the print area (Li et al. 2023). The material (mortar or concrete), usually stored in a silo or container, is pre-mixed with water in a pump that ensures an adequate flow rate and consistent supply to the hose and nozzle. The dimensions of the nozzle determine the layer's thickness and width. The design is defined by a digital model that contains the precise movements and instructions for the 3D printer. Cementitious materials, such as concrete or cement mortar, are carefully prepared to achieve the desired properties for 3D printing. The material is made by mixing binders, aggregates, water, and specific additives to get the right viscosity and flowability. This makes sure that the material flows smoothly through the printing nozzle and that the printed structure stays stable. Various additives, such as plasticizers or rheology modifiers, can be incorporated to enhance the workability and printability of the material. Fibres can also be added to the cement mortar to improve its mechanical properties, such as strength and durability. These are short fibres, typically made of materials like steel, carbon, or polypropylene, that, in the absence of rebars, reinforce the printed structure to prevent cracking or deformation (Pham et al. 2020, van Hierden et al. 2022). The choice of additives and fibres depends on the specific requirements of the printed object, such as its size, shape, and intended use. Overall, the careful selection and incorporation of these materials plays a crucial role in ensuring the success of 3D printing with cement mortar (Boddepalli et al. 2023).

## Properties of 3DP Concrete (3DPC)

Even though printing systems can still benefit from more control and improved hardware, the biggest challenge remains at the material level. The mortar composition plays a crucial role in determining printed structures' strength, durability, and workability. Achieving consistent mortar composition throughout the printing process can be challenging due to variations in material properties, mixing procedures, and environmental conditions (Paritala et al. 2023). Ensuring mix control and quality is essential for producing reliable and structurally sound 3D-printed components. Ensuring adequate pumpability, extrudability and buildability is key to developing a successful 3D printable mix, but to achieve successful printing, contradictory fresh-state requirements must be met (Figure 2) (Rehman and Kim 2021).

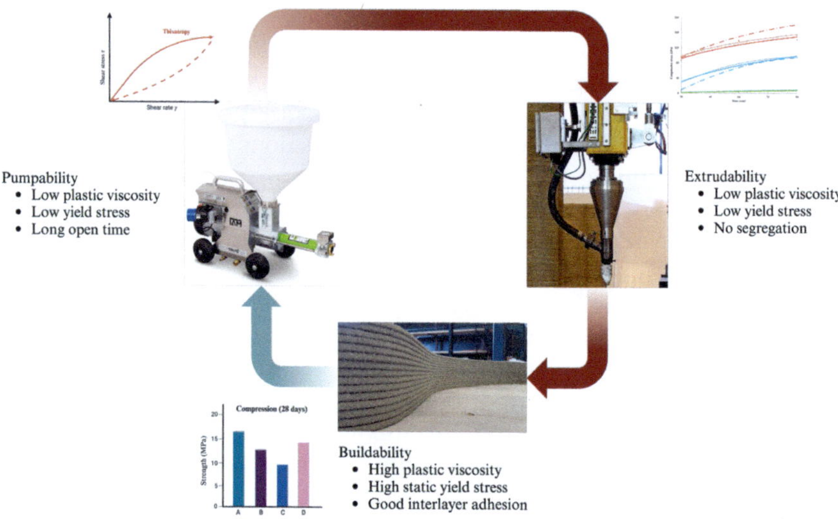



**Figure 2**    Requirements in 3D concrete printing at different printing stages.

## Pumpability and Extrudability

Pumpability can be defined as the ability of the mix to flow, under pressure, through the system up to the printing head without losing its initial properties. However, the extruded material must be stiff enough to ensure that the deposited filaments retain their shape. Pumpability depends on many setup variables, including the hose diameter, pumping distance, and pumping technique (Xiao et al. 2021). A pumpable mixture should have low plastic viscosity and low yield stress. This is necessary to inhibit the migration of fine particles to the surface, as this could cause a lubricating layer on the inner walls of the hose or pipe. The presence of this layer can lead to a final, inhomogeneous product that does not meet the right criteria for printability and increases the risk of blockages in the system.

Extrudability refers to the inherent capacity of the material to undergo unobstructed, seamless extrusion, resulting in a smooth, continuous filament. The composition also needs to be adjusted to the size and shape of the nozzle. Nozzle geometry dictates the quality of the final product and its surface finishing (Jayathilakage et al. 2020).

## Buildability

Buildability is described as the capacity of the printed material to retain its shape and stability when deposited in successive layers. The filament should remain continuous when exiting the nozzle. The fresh state strength needs to be sufficiently high to ensure that the material can be stacked without showing any deformation or buckling (Muthukrishnan et al. 2021).

For the material to be pumpable and extrudable, low plastic viscosity and low dynamic yield stress should be sought, but buildability requires high static yield stress and viscosity. The mortars' properties are also intimately dependent on printing speed, which is directly connected with the interlayer adhesion and the cost-effectiveness of the printing process. All these conflicting properties are to be met in one 'ink'. Therefore, it is not surprising that the biggest task researchers face in 3DPC is to find a suitable, reproducible and scalable mix capable of matching all these properties.

## ADVANTAGES AND CHALLENGES IN 3DCP

One key advantage of 3D concrete printing is its ability to optimise the shape of the structural elements. The ability to create customised and individualised structures is a unique benefit of 3D concrete printing. This opens new possibilities for architectural design and customisation in construction. The use of formwork to prepare concrete is an inhibitor of design freedom because moulds can be expensive, and their use limits the shape of the elements. Printing self-supported material reduces waste generated by discarded moulds, allowing for more flexibility in shape and form. This design flexibility can also be used to optimise the element's shape, reducing material consumption without compromising structural performance (Wang et al. 2023). The precision and customisation offered by the technology can lead to reduced material waste and improved resource efficiency.

3D printing unlocks a new realm of design possibilities, allowing architects and designers to create intricate and unconventional structures. It enables the implementation of organic shapes, intricate detailing, and complex geometries that were previously challenging or impossible to achieve with traditional methods. The ability to customise designs for individual projects enhances design flexibility and fosters innovation within the industry.

In terms of structural behaviour, 3D-printed concrete can exhibit improved performance compared to traditional concrete. While it is known that some process constraints during printing can limit the ductile behaviour of 3D-printed concrete structures, research on the flexural and shear behaviour of 3D printed reinforced concrete beams has shown promising results in terms of strength and performance (Ye et al. 2022). The automation of the construction manufacturing process through digital design and workflow allows for higher degrees of form freedom, adding value to the construction industry.

It provides benefits such as increased productivity, reduced waste, and labour optimisation. Complex, non-standard geometries can be rapidly manufactured. The potential for higher construction speed is crucial for efficient and fast construction, especially in large-scale projects. With 3D printing, construction processes can be significantly expedited (Diggs-McGee et al. 2019). Automated robotic systems can work non-stop, allowing for faster project completion. Increasing speed gains particular relevance in emergency responses or areas where urgent housing is needed. Rapid deployment and on-site construction capabilities make it a valuable tool for disaster relief. In emergencies 3D printers can quickly produce temporary shelters and other essential structures, providing immediate relief to affected communities.

While 3D printing offers significant advantages, several challenges and limitations need to be addressed for its widespread adoption. One area of significant research and development in 3DCP is the use of alternative raw materials in the printing process. Traditional concrete typically relies on Portland cement as the binder, which is known to have a significant carbon footprint. The reliance on Portland cement continued with 3DP because the binder can guarantee good fresh properties and high mechanical strength. However, the lack of coarse aggregates in 3DP mortars means that higher percentages of cement are needed compared with traditional concrete. Most 3DPC compositions today have a high binder concentration and a small aggregate size. Coarse aggregates are rarely employed in printable concrete since most 3D printers use small-size nozzles and hoses. This somehow defeats the aim of using the technology to improve sustainability. To address this issue, researchers are now putting more effort into the use of alternative binders such as slag, metakaolin, and clay as cement replacements. These binders offer several advantages, including reduced carbon emissions, improved durability, and enhanced fire resistance. They can be used as a partial binder replacement for Portland cement or as precursors in alkali-activated materials (Beersaerts et al. 2023a,b).

Issues such as scaling up the technology for larger projects, ensuring structural integrity, and addressing material limitations require further development (Matos et al. 2023). Reinforcement, for example, is critical for structural integrity in construction, but its integration with 3D-printed components poses challenges (Han et al. 2022). The placement of traditional reinforcement materials, such as steel bars or mesh, is difficult to combine with the 3D printing process. Alternative reinforcement strategies, such as fibre reinforcement or embedded reinforcement systems, are being explored, but their effectiveness, long-term durability, and compatibility with 3D printing materials demand much more research (Liu et al. 2022a).

Existing regulations and concrete mix standards do not encompass compositions with fine sand and no coarse aggregates. Even though the term concrete is widely used, 3D-printed concrete falls technically into the category of a mortar. Therefore, the conventional mix design methods commonly employed in traditional concrete are insufficient to guide the mix design of 3D-printed concrete (3DPC), since technically they fall into the mortar category. Despite the existence of several empirical mix designs proposed in recent studies, replicability is a major challenge due to the variability in materials and printing techniques used in 3DPC. Existing codes also do not address the complexities associated with 3D-printed structures; therefore, obtaining the necessary approvals and permits can be challenging. To become a mainstream construction method, standardisation of protocols, materials, and practices is crucial (Siddika et al. 2019).

# 4D PRINTING OF FUNCTIONAL CEMENTITIOUS MATERIALS

Functional cement-based materials, also frequently named 'smart' cement or 'smart' concrete, are mortars or concrete that include in their composition an additive, usually a nanoparticle, which provides extra functionality (Olafusi et al. 2019). The interest in using functional additives (nanoparticles) to modify materials' properties is not recent. It began shortly after the successful development of nanoparticles at a laboratory scale, escalating substantially with the scale-up production and commercial availability of these nanoadditives. Nanoparticles are particles with dimensions on the nanoscale, typically smaller than 100 nm in at least one dimension. Compared to macroscale materials, they exhibit unique properties due to their small size and high surface-to-volume ratio. The much larger surface area compared to macroscale materials results in enhanced reactivity and increased interaction with their surroundings. They exhibit different chemical and physical properties compared to their macroscale counterparts (Singh and Saini 2022).

The main goal of using nanoadditives in cement and concrete is to add value by including an extra property or improving the final product. It can make a common construction material capable of, for example, degrading indoor air pollutants, self-healing microcracks, acting as a sensor for structural damage, and exhibiting improved fresh properties or higher strength (Paul et al. 2018). While research into the application of nanoparticles in construction materials has been prolific, commercial applications remain scarce. The focus has been on increased knowledge and understanding of basic phenomena at the macro- and nanoscale.

The biggest difficulty encountered in the use of these nanoadditives is their effective dispersion. Inhomogeneous distribution leads to uneven properties, increased porosity and a loss of mechanical strength (Xu et al. 2021).

When nanoparticles are combined with 3D concrete printing, they can be precisely integrated into the concrete matrix during the printing process. This integration can be optimised to provide tailored properties, resulting in improvements in the structural, mechanical, and durability characteristics of the printed components (Basha et al. 2023).

## Nano-silica (Fresh Strength)

Nano-$SiO_2$ (nano-silica) accelerates the hydration process by providing nucleation sites for the precipitation of hydration products. This acceleration leads to faster setting times and improved early strength development, enhancing the overall buildability of 3DPC structures. The enhanced pozzolanic reactions and densification of the microstructure contribute to higher compressive and flexural strengths, ensuring the durability and stability of the 3D-printed components. It can increase resistance to water penetration and help regulate calcium leaching, factors that are directly linked to concrete degradation. The effect can be achieved even with very low quantities (0.25%), and its use is more efficient than silica fume (Liu et al. 2023a). Nano-$SiO_2$ serves a dual purpose when added to 3DPC by improving its microstructure and facilitating pozzolanic reactions. The spherical shape of nano-$SiO_2$ particles enhances the flow and workability of 3D-printed concrete (3DPC). This improved workability facilitates smoother pumping and extrusion during the printing process, ensuring a more homogeneous and consistent structure (Mendoza et al. 2019).

## Nano-titania (Self-Cleaning)

Nano-$TiO_2$ (nano-titania) has proved to be highly effective in self-cleaning and depolluting concrete. This nanoparticle does not react directly with cement but acts as a catalyst in photocatalytic reactions. Nano-sized titanium dioxide, in the mineral form anatase, when in contact with humidity in the air, becomes activated when exposed to sunlight. The energy provided by light causes the most external electron on the nano-$TiO_2$ particle to jump from its valence band to the conduction band. This allows the catalyst to promote a reaction between water molecules and common pollutants ($NO_x$, $SO_x$ and VOCs), resulting in inert particles harmless to the environment (Janczarek et al. 2022). This innovative concrete has already been employed in various locations across Europe and Japan. The same photocatalytic process is also known to kill bacteria and mould, making the concrete surface self-cleaning too. This contributes to lower maintenance and improved durability of mortars, especially in highly polluted urban areas.

Research on the use of nano-titania in 3DCP is extremely limited. However, it appears that nano-titania possesses an excellent filling effect that, combined with its high surface area, can contribute to lowering the mortar's porosity and refining its pore structure. For low water-binder ratios, it can improve the mortar's static yield stress, dynamic yield stress and plastic viscosity, which is useful for ensuring good printability (Figure 3) (Liu Q. 2022b).

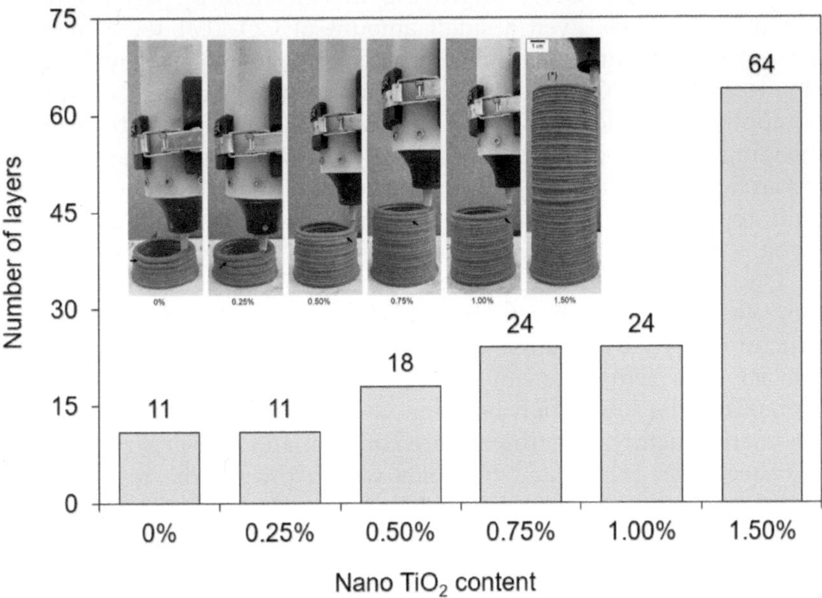


**Figure 3** Failure during buildability test for different nano-TiO$_2$ contents (de Matos et al. 2022). Copyright (2022) Materials, MDPI, Creative Commons CC-BY license.

## Graphene Oxide (Mechanical Strength)

Cementitious matrices undergo complex chemical reactions that lead to complex products with irregular structures. This produces a disordered framework, potentially containing microcracks and pores (Scrivener et al. 2023). Reinforcement materials such as steel bars, metal, and polymeric fibres have been applied to mitigate these issues and improve stability and performance. While these fibres can effectively prevent the propagation of cracks at the macroscale level, they are usually ineffective for microcracks. Nanomaterials are more effective at preventing nano and microcrack formation, mostly because, due to their very high surface area, they allow for more interaction with the cement matrix (Sun et al. 2023).

Graphene oxide (GO), a highly oxidised derivative of graphene (a single layer of graphite), is a two-dimensional (2-D) nanomaterial with significant functional groups of oxygen molecules. It is the main choice for cementitious matrices, showing good attachment with separate oxygen functional groups and exhibiting strong reactivity with cement. GO is highly covalent in the carboxylic groups on the boundaries. These functional groups react with cement hydration products such as C–S–H and Ca $(OH)_2$, improving the mechanical performance of cement composites. Even a small amount of GO (0.01 wt%) increases the compressive and flexural strengths of cement paste by 25% and 41%, respectively. GO has been used in functional cement composites for applications such as self-sensing, self-healing, and electromagnetic shielding. Cement composites with self-sensing properties are a strong research topic in the field of functional cement materials, and they have been applied for traffic monitoring and structural health monitoring (Long 2023).

One of the primary purposes of incorporating GO in 3D-printed concrete is to enhance the mechanical strength and durability of the final structure. GO, due to its high surface area and strong bonding capabilities, can act as a reinforcement agent, creating a stronger cementitious composite. The interaction between GO and the cement mortar alters the microstructure, forming a more compact and denser structure. The GO sheets can bridge and fill voids within the matrix, resulting in a reduction of microcracks and enhanced mechanical properties. This improved microstructure can ultimately contribute to increased tensile and compressive strength, enhancing the overall performance of the 3D-printed concrete. The unique mechanical properties of GO, including its exceptional tensile strength and elasticity, enable it to effectively disperse stress and enhance the material's ability to withstand bending forces. This dispersion of stress can help prevent crack initiation and propagation within the concrete, ultimately enhancing its flexural properties. The toughening effect achieved by the incorporation of GO can improve the structural integrity (Figure 4) and longevity of the 3D-printed concrete components (Basha et al. 2023).

Despite the promising advantages of integrating GO into 3D-printed concrete, several challenges need to be addressed for successful implementation. One significant issue is achieving a homogeneous dispersion of GO within the cementitious matrix. Due to its tendency to agglomerate, reaching a uniform distribution of GO throughout the concrete can be difficult, potentially limiting its benefits. Proper dispersion methods and techniques need to be developed to ensure a consistent and effective distribution of GO within the mortar. While GO can also be employed to increase electrical conductivity, this can only be achieved with good dispersion in the cement matrix (Lu et al. 2018). A better alternative, when the focus is on self-sensing, is the use of carbon nanotubes (CNTs).

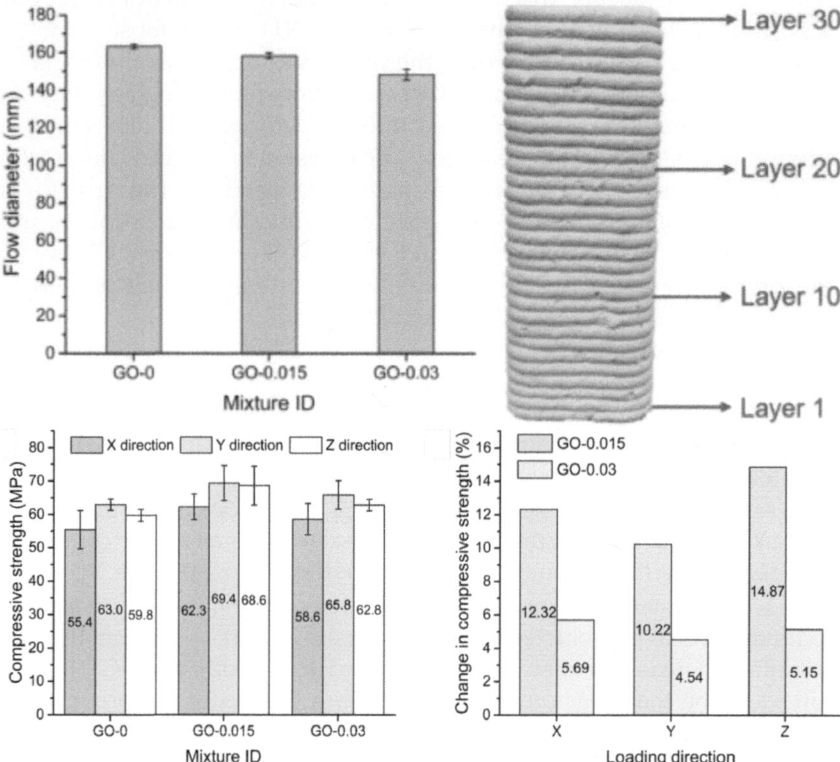


**Figure 4:** Flow diameter of different mixtures and front view of printed hollow structures with a GO mixture (top). 28-day compressive strengths of printed specimens and change of compressive strength (bottom) (Liu et al. 2023b). Copyright (2023) Addit. Manuf. Lett., Elsevier, Creative Commons CC-BY license.

## Carbon Nanotubes (Self-Sensing)

CNTs are being explored as a solution to provide self-sensing capabilities to 3D-printed concrete since, in the absence of reinforcement, structural health monitoring is especially important. Due to their electrical conductivity, CNTs can act as sensors within the printed structure, detecting stress, strain, or other mechanical deformations. These sensors can provide real-time data on the structural integrity of the printed component, allowing for timely maintenance or repairs to ensure long-term durability and safety (Wang and Aslani 2023).

CNTs are remarkable nanomaterials with a cylindrical nanostructure composed of rolled-up graphene sheets. These cylindrical tubes can have single or multiple layers of graphene and exhibit exceptional mechanical, electrical, and thermal properties. CNTs are typically produced through processes such as arc discharge, laser ablation, or chemical vapour deposition (CVD). Among these methods, CVD is the most common and allows for better control over the properties of the resulting CNTs. In the production of CNTs using CVD, a carbon-containing gas, such as acetylene or methane, is introduced into a reactor at high temperatures (around 600–1000°C). The presence of a catalyst, usually transition metal nanoparticles such as iron, nickel, or cobalt, facilitates the growth of carbon nanotubes by acting as a site for carbon atom adsorption and subsequent nanotube nucleation and growth. The resulting CNTs can vary in diameter, length, and chirality, which influences their properties and potential applications (Gupta et al. 2019).

In cement-based mortars and concrete, CNTs are incorporated to enhance the material's mechanical strength, electrical conductivity, and durability. CNTs can be dispersed within the cement matrix using various methods, including sonication, mechanical mixing, or surfactant-assisted dispersion. Incorporating CNTs into 3D-printed concrete comes with specific challenges. Achieving a homogeneous dispersion of CNTs within the concrete mixture is essential to ensuring consistent material properties throughout the printed structure (Figure 5). CNTs tend to agglomerate, requiring careful consideration of the dispersion technique to avoid clustering (Dulaj et al. 2022a). Additionally, the potential increase in material cost due to the addition of CNTs is a consideration that needs to be balanced with the desired improvements in performance. The impact of CNTs on the rheology, buildability, and printability of 3D-printed concrete is substantial. The addition of CNTs can modify the material's rheological behaviour, affecting its flow properties and extrudability during the printing process. To maintain suitable rheology and enable accurate layer-by-layer deposition of the material, proper optimisation of CNT content and dispersion techniques are required. CNTs enhance the buildability of 3D-printed concrete by improving interlayer adhesion and reducing the risk of cracking during printing and curing. Their high aspect ratio and reinforcing properties contribute to a more cohesive printed component (Figure 5), resulting in higher mechanical strength and overall stability (Dulaj et al. 2022b; Sun et al. 2020).

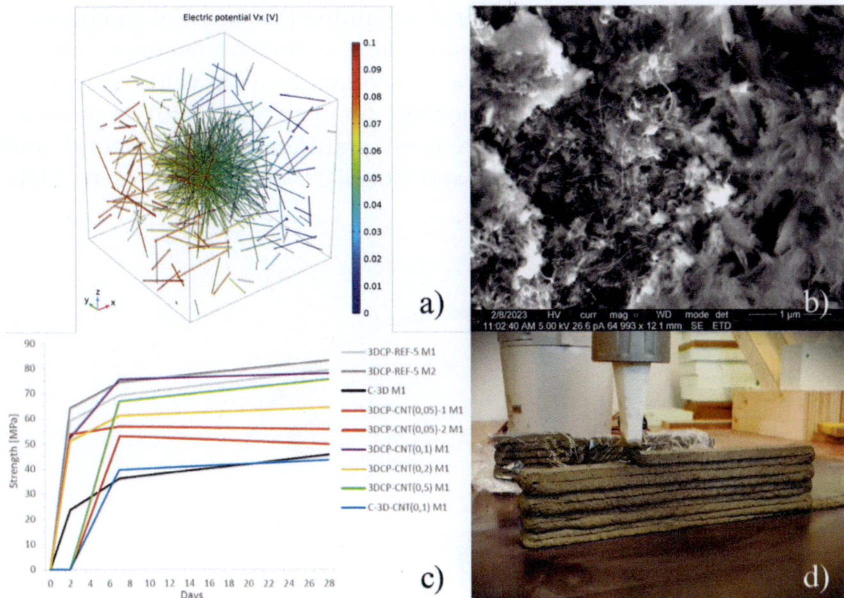



**Figure 5** Distribution of CNTs in 3DCP (a), SEM image of 3DP CNT-cement composite (b), compression strength for different wt% of CNTs (c) and sample of 3DP CNT-cement composite (d) (Dulaj et al. 2022b, Suijs 2023).

Nanomaterials are at the forefront of innovation in the construction industry and have opened a world of possibilities, from improving structural properties to enhancing sustainability and reducing maintenance needs. The marriage of functionality and 3D printing is a promising avenue for the future of construction, one where buildings and infrastructure are not just static structures but dynamic, responsive systems that can adapt and self-repair. Nonetheless, it is essential to emphasise that while the potential of nanomaterials in construction is immense, challenges such as finding optimal compositions, dispersion, cost, and scaling up to industrial production must be addressed (Utsev et al. 2022).

# FROM 4D TO 5D PRINTING

Research on 3DCP is still heavily reliant on experimental investigations, and the development of new compositions is based on trial and error. Computational simulations remain limited and focus more on the macroscale behaviour of the printed material. Cement-based compositions are particularly challenging because they are multi-size, multiphase, and evolve over time. Even the most advanced computational models fail when it comes to fully understanding and explaining the chemical and physical transformations occurring in these mortars at different temporal and size scales. Although current numerical methods provide valuable insights into the performance of 3DP concrete, for their successful application, these modelling techniques require significant experimental and computational resources (Khan and Koç 2022, Nguyen-Van et al. 2021).

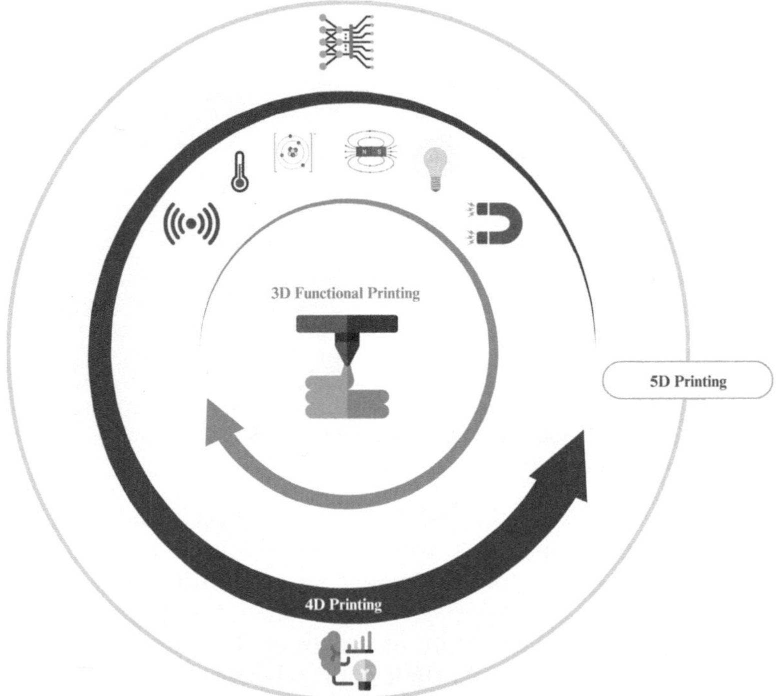


**Figure 6**   5D printing of functional materials

Simulations are only as good as the real-world data they are based on. Hence, significant efforts are invested in collecting and verifying experimental data, which serves as the foundation for these simulations. This validation process is essential in guaranteeing that the models accurately represent the behaviour of the materials and processes involved in 3DCP. To validate simulations and improve their accuracy, high-resolution imaging and in-line testing are often employed, allowing researchers to capture detailed information about the printing process, the interaction between layers, and the material properties at different stages. The computational power required for running simulations is significant. 3DCP models can demand substantial computing resources for performing complex simulations efficiently. The development and refinement of numerical models is an iterative process. It often requires multiple runs of simulations to optimise and adjust parameters. This iteration process demands both computational resources and time for analysis and interpretation of results. Machine learning has the potential to replace or complement traditional computational modelling, helping to uncover patterns, relationships, and dependencies that might be challenging to represent explicitly in equations.

Machine learning is a subfield of artificial intelligence that involves the development of algorithms and statistical models, enabling computer systems to automatically learn and improve their performance of a specific task from experience or data. It covers the study of pattern recognition, data analysis, and optimisation techniques, and it aims to enable machines to make predictions, classifications, or decisions by identifying and generalising patterns within data, thereby enhancing their ability to adapt and evolve as new information becomes available. The use of ML in materials science can revolutionise the way materials are discovered, designed, and optimised. These algorithms can help identify novel materials with desirable properties by analysing vast databases of existing materials and their characteristics. By recognising patterns in material data, they can suggest new compositions or structures for specific applications (Hsu et al. 2022, Yang et al. 2021). ML models can predict various material properties, such as electrical conductivity, thermal conductivity, and mechanical strength, based on the material's composition and structure. This is invaluable for designing materials for specific functions and optimising existing ones (Hamel et al. 2019). Machine learning techniques can assist in the analysis of experimental data from techniques like X-ray diffraction, spectroscopy, and electron microscopy and help researchers interpret complex data, identify structural features, and detect defects in materials. It can optimise experiments and testing procedures, reducing the time and resources required to evaluate materials, and be used to predict material degradation, wear and tear, or potential failure in structural materials. This is crucial for maintaining the safety and integrity of critical infrastructure. Machine learning can aid in the design of "smart" materials that can respond to external stimuli. For example, shape-memory alloys or self-healing materials can be optimised with the help of machine learning algorithms. ML models provide 4D materials with a fifth dimension, which, combined with additive manufacturing, can be defined as 5D printing (Figure 6) (Faruque et al. 2023, Pugliese and Regondi 2022).

## Machine Learning in 3D Concrete Printing

The utilisation of machine learning in 3DCP represents a new stage and a growing opportunity to streamline and enhance the process, potentially reducing resource demands associated with traditional numerical methods. Examples of applications in the field are still scarce, but the possibilities are clear even at this stage, ranging from property prediction to structural optimisation (Goh et al. 2021).

## Mix Optimisation

At the core of 3DCP is the material itself. The mortar's composition, including the choice of aggregates, admixtures, and additives, significantly impacts the properties of the printed structure. Traditional methods of material development rely on a lengthy trial-and-error process where different combinations are tested in the lab and then iteratively optimised for the printer, often resulting in a protracted and costly experimentation phase. The high variability of cement types and other raw materials makes most studies on mix development difficult to reproduce. Quite often, they do not outline the design method or the parameters that were considered, but experience shows that the number of tests to achieve a printable mix is extremely high. Machine learning can transform this process by enabling the creation of predictive models that identify optimal material compositions (Ziolkowski and Niedostatkiewicz 2019). These models can consider a multitude of variables, including the type and proportion of cementitious materials, aggregates, and nanoparticles. By feeding the algorithm with historical data (rheology, fresh strength, etc.), the model can recommend precise material compositions that meet the desired properties of the final structure, such as compressive strength, durability, and functionality (Ali et al. 2023, Kekez and Kubica 2020).

Work on the use of ML to assist in the development of 3D printable cement and concrete is extremely limited. A viable strategy seems to be to use factorial design to predetermine a set of optimal combinations of raw materials, conduct a preselected group of tests to gather sufficient data to feed the algorithm, and, through multiple iterations, find new compositions. The use of ML is not limited to optimising compositions but also to assisting in the discovery of more sustainable compositions with lower cement content (Gandhi and Hasan 2022).

## Print Path Optimisation

Machine learning can be applied to print path optimisation in 3D concrete printing. By analysing the design of a structure, material properties, and other factors, ML tools can determine the most efficient and precise way to deposit the mortar layers. The goal is to reduce waste, save time, and minimise energy consumption while ensuring that the printed structure meets design specifications. The 3D design of a concrete structure can be analysed to establish where the most critical load-bearing areas are. The model can then plan the print path for optimal structural integrity. An optimised print path can substantially reduce the time required to print a structure and minimise the amount of material required. Since, quite often, the geometries of 3DCP are extremely complex, ML tools can be used to minimise unnecessary movements (Ma et al. 2022).

## Real-time Quality Control

Machine learning can be used to implement real-time quality control during the printing process. Computer vision algorithms can analyse each layer as it is printed, detecting defects, voids, or inconsistencies in the structure. The system can then make immediate adjustments to correct any issues. By detecting defects and inconsistencies in the printed layers and applying immediate corrective measures, it is possible to eliminate the need for post-processing. Computer vision algorithms can analyse images and videos of the printing process and recognise irregularities in real-time, detecting defects, voids, and inconsistencies (Ma et al. 2020). When sensors are embedded in the 3D printer, real-time data on variables like material temperature, humidity, flow rate, and pressure can be fed to the algorithms. The system can process this data to identify anomalies or deviations such as uneven layer thickness, material flow problems, or the beginning of nozzle clogs. Combining cameras with temperature, humidity, and pressure sensors will provide a comprehensive view of the printing process and material conditions. As sensor technology continues to improve, the quality and quantity of real-time data available will increase, allowing for more precise control and monitoring (Baduge et al. 2022).

## Limitations and Further Research

One limitation of using ML in 5D concrete printing (5DCP) is the availability of high-quality and diverse training data. ML algorithms require extensive training on reliable and diverse datasets to effectively learn and make accurate predictions. Advanced data collection and processing capabilities, as well as compatibility with the printing equipment and control systems, can make the integration into existing 3D concrete printing systems complex. Interpretation and explanation of ML predictions in 3D concrete printing is another important challenge. ML algorithms often act as black boxes, making predictions without providing a clear understanding of the underlying factors that led to them. Recent advancements in explainable AI are expected to enhance the transparency and trustworthiness of ML predictions, facilitating their adoption in 5DCP.

## CONCLUSION

In this chapter, we analysed the opportunities for combining machine learning and 3D printing in the development of functional cement-based materials. ML offers not only opportunities to optimise the printing process but also to assist in the discovery and optimisation of cement-based materials. Functional construction materials require the addition of special additives (nanoparticles) to the printable mix. ML tools can substantially reduce the experimentation phase, assist in mix development, and provide real-time control, making 4D printing a reality for cement-based materials.

## ACKNOWLEDGEMENTS:

This work was partially supported by the project "Additive manufacturing of functional construction materials on-demand" (with project number 17895) of the research programme "Materialen NL: Challenges 2018," which is financed by the Dutch Research Council (NWO).

## REFERENCES

Ahmed, G.H. 2023. A review of "3D concrete printing": Materials and process characterization, economic considerations and environmental sustainability. J. Build. Eng. 66: 105863. https://doi.org/10.1016/j.jobe.2023.105863.

Ali, A., Riaz, R.D., Malik, U.J., Abbas, S.B., Usman, M., Shah, M.U., et al. 2023. Machine learning-based predictive model for tensile and flexural strength of 3D-printed concrete. Materials. 16(11): 4149. https://doi.org/10.3390/MA 16114149/S1.

Altıparmak, S.C., Yardley, V.A., Shi, Z. and Lin, J. 2022. Extrusion-based additive manufacturing technologies: State of the art and future perspectives. J. Manuf. Processes. 83: 607–636. https://doi.org/10.1016/J.JMAPRO.2022.09.032.

Alves, J.L., Santana, L. and Rangel, B. 2023. 4D printing and construction: Reality, future, or science fiction? pp. 155–175. *In*: Rangel, B., Guimarães, A.S., Lino, J., Santana, L. (eds). 3D Printing for Construction with Alternative Materials. Digital Innovations in Architecture, Engineering and Construction. Springer, Cham. https://doi.org/10.1007/978-3-031-09319-7_7.

Baduge, S.K., Thilakarathna, S., Perera, J.S., Arashpour, M., Sharafi, P., Teodosio, B., et al. 2022. Artificial intelligence and smart vision for building and construction 4.0: Machine and deep learning methods and applications. Autom. Constr. 141: 104440. https://doi.org/10.1016/J.AUTCON.2022.104440.

Beersaerts, G., Hertel, T., Lucas, S. and Pontikes, Y. 2023a. Promoting the use of Fe-rich slag in construction: Development of a hybrid binder for 3D printing.

Cem. Concr. Compos. 138: 104959. https://doi.org/10.1016/j.cemconcomp.20 23.104959.

Beersaerts, G., Soete, J., Giels, M., Eykens, L., Lucas, S. and Pontikes, Y. 2023b. 3D printing of an iron-rich slag based hybrid mortar. A durable, sustainable and cost-competitive product? Cem. Concr. Compos. 144: 105304. https://doi. org/10.1016/j.cemconcomp.2023.105304.

Boddepalli, U., Panda, B. and Ranjani Gandhi, I.S. 2023. Rheology and printability of Portland cement based materials: a review. J. Sustainable Cem.-Based Mater. 12(7): 789–807. https://doi.org/10.1080/21650373.2022.2119620.

Chang, Z., Chen, Y., Schlangen, E. and Šavija, B. 2023. A review of methods on buildability quantification of extrusion-based 3D concrete printing: From analytical modelling to numerical simulation. Dev. Built Environ. 16: 100241. https://doi.org/10.1016/J.DIBE.2023.100241.

de Matos, P., Zat, T., Corazza, K., Fensterseifer, E., Sakata, R., Mohamad, G., et al. 2022. Effect of $TiO_2$ nanoparticles on the fresh performance of 3D-Printed cementitious materials. Materials. 15(11): 3896. https://doi.org/10.3390/MA15 113896/S1.

Diggs-McGee, B., Kreiger, E., Kreiger, M. and Case, M. 2019. Print time vs. elapsed time: A temporal analysis of a continuous printing operation for additive constructed concrete. Addit. Manuf. 28: 205–214. https://doi.org/10.1016/j. addma.2019.04.008.

Dulaj, A., Salet, T.A.M. and Lucas, S.S. 2022a. Mechanical properties and self-sensing ability of graphene-mortar compositions with different water content for 3D printing applications. Mater. Today Proc. 70: 412–417. https:// doi.org/10.1016/J.MATPR.2022.09.278

Dulaj, A., Suijs, M.P.M., Salet, T.A.M. and Lucas, S.S. 2022b. Incorporation and characterization of multi-walled carbon nanotube concrete composites for 3D printing applications. pp. 119–125. *In*: Buswell, R., Blanco, A., Cavalaro, S. and Kinnell, P. (eds). Third RILEM International Conference on Concrete and Digital Fabrication. DC 2022. RILEM Bookseries, vol 37. Springer, Cham. https://doi.org/10.1007/978-3-031-06116-5_18.

Faruque, M.O., Lee, Y., Wyckoff, G.J. and Lee, C.H. 2023. Application of 4D printing and AI to cardiovascular devices. J. Drug Delivery Sci. Technol. 80: 104162. https://doi.org/10.1016/J.JDDST.2023.104162.

Gandhi, A. and Hasan, M.M.F. 2022. Machine learning for the design and discovery of zeolites and porous crystalline materials. Curr. Opin. Chem. Eng. 35: 100739. https://doi.org/10.1016/J.COCHE.2021.100739.

Goh, G.D., Sing, S.L. and Yeong, W.Y. 2021. A review on machine learning in 3D printing: Applications, potential, and challenges. Artif. Intell. Rev. 54(1): 63–94. https://doi.org/10.1007/S10462-020-09876-9/FIGURES/13.

Guimarães, A.S., Delgado, J.M.P.Q. and Lucas, S.S. 2021. Additive manufacturing on building construction. DDF. 412: 207–216. https://doi.org/10.4028/WWW. SCIENTIFIC.NET/DDF.412.207.

Gupta, N., Gupta, S.M. and Sharma, S.K. 2019. Carbon nanotubes: Synthesis, properties and engineering applications. Carbon Lett. 29(5): 419–447. https:// doi.org/10.1007/S42823-019-00068-2/FIGURES/13.

Hamel, C.M., Roach, D.J., Long, K.N., Demoly, F., Dunn, M.L. and Qi, H.J. 2019. Machine-learning based design of active composite structures for 4D printing. Smart Mater. Struct. 28(6): 65005. https://doi.org/10.1088/1361-665X/AB1439.

Han, X., Yan, J., Liu, M., Huo, L. and Li, J. 2022. Experimental study on large-scale 3D printed concrete walls under axial compression. Autom. Constr. 133: 103993. https://doi.org/10.1016/J.AUTCON.2021.103993.

Heinz, O. and Heinz, H. 2021. Cement interfaces: Current understanding, challenges, and opportunities. Langmuir, 37(21): 6347–6356. https://doi.org/10.1021/acs.langmuir.1c00617.

Hsu, Y.C., Yang, Z. and Buehler, M.J. 2022. Generative design, manufacturing, and molecular modeling of 3D architected materials based on natural language input. APL Materials. 10(4): 41107. https://doi.org/10.1063/5.0082338/2835002.

Huang, Z., Shao, G. and Li, L. 2023. Micro/nano functional devices fabricated by additive manufacturing. Prog. Mater Sci. 131: 101020. https://doi.org/10.1016/J.PMATSCI.2022.101020.

Inayath Basha, S., Ur Rehman, A., Aziz, M.A. and Kim, J.H. 2023. Cement composites with carbon-based nanomaterials for 3D concrete printing applications – A review. Chem. Rec. 23(4): e202200293. https://doi.org/10.1002/TCR.202200293.

Janczarek, M., Klapiszewski, Ł., Jędrzejczak, P., Klapiszewska, I., Ślosarczyk, A. and Jesionowski, T. 2022. Progress of functionalized $TiO_2$-based nano-materials in the construction industry: A comprehensive review. Chem. Eng. J. 430: 132062. https://doi.org/10.1016/j.cej.2021.132062.

Jayathilakage, R., Sanjayan, J. and Rajeev, P. 2020. Characterizing extrudability for 3D concrete printing using discrete element simulations. RILEM Bookseries. 28: 290–300. https://doi.org/10.1007/978-3-030-49916-7_30/FIGURES/5.

Kekez, S. and Kubica, J. 2020. Connecting concrete technology and machine learning: proposal for application of ANNs and CNT/concrete composites in structural health monitoring. RSC Adv. 10(39): 23038–23048. https://doi.org/10.1039/D0RA03450A.

Khan, S.A. and Koç, M. 2022. Numerical modelling and simulation for extrusion-based 3D concrete printing: The underlying physics, potential, and challenges. Results Mater. 16: 100337. https://doi.org/10.1016/J.RINMA.2022.100337

Li, S., Lan, T., Mendis, P. and Tran, P. 2023. Robotics in 3D concrete printing: Current progress & challenges. pp. 27–42. *In*: Noroozinejad F.E., Noori, M., Yang, T.T.Y., Lourenço, P.B., Gardoni, P., Takewaki, I., et al. (eds). Automation in Construction toward Resilience: Robotics, Smart Materials and Intelligent Systems (1st ed.). CRC Press. https://doi.org/10.1201/9781003325246.

Liu, B., Liu, X., Li, G., Geng, S., Li, Z., Weng, Y., et al. 2022a. Study on anisotropy of 3D printing PVA fiber reinforced concrete using destructive and non-destructive testing methods. Case Stud. Constr. Mater. 17: e01519. https://doi.org/10.1016/J.CSCM.2022.E01519.

Liu, Q., Jiang, Q., Huang, M., Xin, J. and Chen, P. 2022b. The fresh and hardened properties of 3D printing cement-base materials with self-cleaning nano-$TiO_2$:An exploratory study. J. Cleaner Prod. 379: 134804. https://doi.org/10.1016/j.jclepro.2022.134804.

Liu, B., Zhou, H., Meng, H., Pan, G. and Li, D. 2023a. Fresh properties, rheological behavior and structural evolution of cement pastes optimized using highly dispersed in situ controllably grown Nano-SiO$_2$. Cem. Concr. Compos. 135: 104828. https://doi.org/10.1016/J.CEMCONCOMP.2022.104828.

Liu, J., Tran, P., Ginigaddara, T. and Mendis, P. 2023b. Exploration of using graphene oxide for strength enhancement of 3D-printed cementitious mortar. Addit. Manuf. Lett. 7: 100157. https://doi.org/10.1016/J.ADDLET.2023.100157

Long, W.J. 2023. Design, performance, and mechanism of cement-based materials with 2D nanomaterials. pp. 127–159. *In*: Khayat, K.H. and Meng, W. (eds). Nanotechnology for Civil Infrastructure: Innovation and Eco-Efficiency of Nanostructured Cement-Based Materials, Series: Micro and Nano Technologies. Elsevier. https://doi.org/10.1016/B978-0-12-817832-4.00001-8.

Lu, Z., Hanif, A., Sun, G., Liang, R., Parthasarathy, P. and Li, Z. 2018. Highly dispersed graphene oxide electrodeposited carbon fiber reinforced cement-based materials with enhanced mechanical properties. Cem. Concr. Compos. 87: 220–228. https://doi.org/10.1016/j.cemconcomp.2018.01.006.

Lucas, S.S., Moxham, C., Tziviloglou, E. and Jonkers, H. 2018. Study of self-healing properties in concrete with bacteria encapsulated in expanded clay. Sci. Technol. Mater. 30: 93–98. https://doi.org/10.1016/j.stmat.2018.11.006.

Lucas, S.S. and Barroso de Aguiar, J.L. 2019. Evaluation of latent heat storage in mortars containing microencapsulated paraffin waxes – A selection of optimal composition and binders. Heat Mass Transfer. 55(9): 2429–2435. https://doi.org/10.1007/s00231-019-02594-1.

Ma, G., Li, Y., Wang, L., Zhang, J. and Li, Z. 2020. Real-time quantification of fresh and hardened mechanical property for 3D printing material by intellectualization with piezoelectric transducers. Constr. Build. Mater. 241: 117982. https://doi.org/10.1016/j.conbuildmat.2019.117982.

Ma, Z., Wan, W., Song, L., Liu, C., Liu, H. and Wu, Y. 2022. An approach of path optimization algorithm for 3D concrete printing based on graph theory. Appl. Sci. 12(22): 11315. https://doi.org/10.3390/app122211315.

Matos, A.M., Emiroğlu, M., Subasi, S., Marasli, M., Guimrães, A.S. and Delgado, J. 2023. Architectonic cement-based composites 3D printing. pp. 67–89. *In*: 3D Printing for Construction with Alternative Materials. Springer, Cham. https://doi.org/10.1007/978-3-031-09319-7_4.

Mendoza Reales, O.A., Duda, P., Silva, E.C.C.M., Paiva, M.D.M. and Filho, R.D.T. 2019. Nanosilica particles as structural buildup agents for 3D printing with Portland cement pastes. Constr. Build. Mater. 219: 91–100. https://doi.org/10.1016/j.conbuildmat.2019.05.174.

Milazzo, M. and Libonati, F. 2022. The synergistic role of additive manufacturing and artificial intelligence for the design of new advanced intelligent systems. Adv. Intell. Syst. 4(6): 2100278. https://doi.org/10.1002/AISY.202100278.

Muthukrishnan, S., Ramakrishnan, S. and Sanjayan, J. 2021. Technologies for improving buildability in 3D concrete printing. Cem. Concr. Compos. 122: 104144. https://doi.org/10.1016/J.CEMCONCOMP.2021.104144.

Nguyen-Van, V., Panda, B., Zhang, G., Nguyen-Xuan, H. and Tran, P. 2021. Digital design computing and modelling for 3-D concrete printing. Autom. Constr. 123: 103529. https://doi.org/10.1016/J.AUTCON.2020.103529.

Olafusi, O.S., Sadiku, E.R., Snyman, J., Ndambuki, J.M. and Kupolati, W.K. 2019. Application of nanotechnology in concrete and supplementary cementitious materials: a review for sustainable construction. SN Appl. Sci. 1(6): 1–8. https://doi.org/10.1007/s42452-019-0600-7.

Paritala, S., Singaram, K.K., Bathina, I., Khan, M.A. and Jyosyula, S.K.R. 2023. Rheology and pumpability of mix suitable for extrusion-based concrete 3D printing – A review. Constr. Build. Mater. 402: 132962. https://doi.org/10.1016/J.CONBUILDMAT.2023.132962.

Paul, S.C., van Rooyen, A.S., van Zijl, G.P.A.G. and Petrik, L.F. 2018. Properties of cement-based composites using nanoparticles: A comprehensive review. Constr. Build. Mater. 189: 1019–1034. https://doi.org/10.1016/j.conbuildmat.2018.09.062.

Pham, L., Tran, P. and Sanjayan, J. 2020. Steel fibres reinforced 3D printed concrete: Influence of fibre sizes on mechanical performance. Constr. Build. Mater. 250: 118785. https://doi.org/10.1016/j.conbuildmat.2020.118785.

Pugliese, R. and Regondi, S. 2022. Artificial intelligence-empowered 3D and 4D printing technologies toward smarter biomedical materials and approaches. Polymers. 14(14): 2794. https://doi.org/10.3390/POLYM14142794.

Rehman, A.U. and Kim, J.H. 2021. 3D concrete printing: A systematic review of rheology, mix designs, mechanical, microstructural, and durability characteristics. Materials. 14(14): 3800. https://doi.org/10.3390/MA14143800.

Scrivener, K.L., Matschei, T., Georget, F., Juilland, P. and Mohamed, A.K. 2023. Advances in hydration and thermodynamics of cementitious systems. Cem. Concr. Res. 174: 107332. https://doi.org/10.1016/J.CEMCONRES.2023.107332.

Siddika, A., Mamun, Md. A. Al, Ferdous, W., Saha, A.K. and Alyousef, R. 2019. 3D-printed concrete: Applications, performance, and challenges. J. Sustainable Cem.-Based Mater. https://doi.org/10.1080/21650373.2019.1705199.

Singh, G. and Saini, B. 2022. Nanomaterial in cement industry: A brief review. Innovative Infrastruct. Solutions. 7(1): 1–13. https://doi.org/10.1007/S41062-021-00649-Z/TABLES/4.

Suijs, M.P.M. 2023. Mechanical and conductive properties of 3D printed concrete with added carbon nanotubes [Master, Eindhoven University of Technology]. Student thesis: Master. https://research.tue.nl/en/studentTheses/mechanical-and-conductive-properties-of-3d-printed-concrete-with-

Sun, X., Wang, Q., Wang, H. and Chen, L. 2020. Influence of multi-walled nanotubes on the fresh and hardened properties of a 3D printing PVA mortar ink. Constr. Build. Mater. 247: 118590. https://doi.org/10.1016/J.CONBUILDMAT.2020.118590.

Sun, H., Zou, H., Ren, J., Xu, G. and Xing, F. 2023. Synthesis of a novel graphene oxide/belite cement composite and its effects on flexural strength and interfacial transition zone of ordinary portland cement mortars. Constr. Build. Mater. 402: 133009. https://doi.org/10.1016/J.CONBUILDMAT.2023.133009.

Utsev, T., Tiza, T.M., Mogbo, O., Kumar Singh, S., Chakravarti, A., Shaik, N., et al. 2022. Application of nanomaterials in civil engineering. Mater. Today Proc. 62: 5140–5146. https://doi.org/10.1016/J.MATPR.2022.02.480.

van Hierden, N.Z., Gauvin, F., Lucas, S.S., Salet, T.A.M. and Brouwers, H.J.H. 2022. Use of hemp fibres in 3D printed concrete. Construction Technologies and Architecture. 1: 758–765. https://doi.org/10.4028/WWW.SCIENTIFIC.NET/CTA.1.758.

Wang, L. and Aslani, F. 2023. Structural performance of reinforced concrete beams with 3D printed cement-based sensor embedded and self-sensing cementitious composites. Eng. Struct. 275: 115266. https://doi.org/10.1016/J.ENGSTRUCT.2022.115266.

Wang, L., Ye, K., Wan, Q., Li, Z. and Ma, G. 2023. Inclined 3D concrete printing: Build-up prediction and early-age performance optimization. Addit. Manuf. 71: 103595. https://doi.org/10.1016/J.ADDMA.2023.103595.

Xiao, J., Ji, G., Zhang, Y., Ma, G., Mechtcherine, V., Pan, J., et al. 2021. Large-scale 3D printing concrete technology: Current status and future opportunities. Cem. Concr. Compos. 122: 104115. https://doi.org/10.1016/J.CEMCONCOMP.2021.104115.

Xu, Z., Huang, Z., Liu, C., Deng, H., Deng, X., Hui, D., et al. 2021. Research progress on key problems of nanomaterials-modified geopolymer concrete. Nanotechnol. Rev. 10(1): 779–792. https://doi.org/10.1515/NTREV-2021-0056/ASSET/GRAPHIC/J_NTREV-2021-0056_FIG_009.JPG.

Xue, D., Balachandran, P.V, Hogden, J., Theiler, J., Xue, D. and Lookman, T. 2016. Accelerated search for materials with targeted properties by adaptive design. Nat. Commun. 7: 11241. https://doi.org/10.1038/ncomms11241.

Yang, Z., Yu, C.H. and Buehler, M.J. 2021. Deep learning model to predict complex stress and strain fields in hierarchical composites. Sci. Adv. 7(15): 7416. https://www.science.org/doi/10.1126/sciadv.abd7416.

Ye, J., Weng, Y., Du, H., Li, M., Yu, J. and Nasir Uddin, M. 2022. Feasibility of using ultra-high ductile concrete to print self-reinforced hollow structures. RILEM Bookseries. 37: 133–138. https://doi.org/10.1007/978-3-031-06116-5_20/FIGURES/6.

Zaki, M., Sharma, S., Gurjar, S.K., Goyal, R., Jayadeva and Krishnan, N.M.A. 2022. Cementron: Machine learning the constituent phases in cement clinker from optical images. https://arxiv.org/abs/2211.03223v1.

Zhang, H., Xu, Y., Gan, Y., Chang, Z., Schlangen, E. and Šavija, B. 2020. Microstructure informed micromechanical modelling of hydrated cement paste: Techniques and challenges. Constr. Build. Mater. 251: 118983. https://doi.org/10.1016/J.CONBUILDMAT.2020.118983.

Ziolkowski, P. and Niedostatkiewicz, M. 2019. Machine learning techniques in concrete mix design. Materials. 12(8): 1256. https://doi.org/10.3390/ma1208 1256.

Chapter **4**

# 3D-Printed Mortars with Marble Powder Towards Sustainable Construction


**Manuel Jesus[1]** (https://orcid.org/0000-0002-8998-7205)
**Elis Ribeiro[1]** (https://orcid.org/0009-0002-5795-0229)
**João Teixeira*,[1]** (https://orcid.org/0000-0002-3359-5157)
**Bárbara Rangel[1,2]** (https://orcid.org/0000-0002-5911-9423)
**Ana S. Guimarães[1]** (https://orcid.org/0000-0002-8467-6264)
**and Jorge L. Alves[3]** (https://orcid.org/0000-0002-9327-9092)

[1]CONSTRUCT, Faculty of Engineering, University of Porto, Porto, Portugal
[1,2]CEAU, Faculty of Architecture of University of Porto, Porto, Portugal
[3]INEGI, Faculty of Engineering, University of Porto, Porto, Portugal


## INTRODUCTION

3D Construction Printing (3DCP) represents a transformative approach to modern Construction, offering unparalleled design freedom, reduced material waste, and increased production speed. In this innovative landscape, the integration of sustainable materials into 3D-printed


*For Correspondence: jhildtex@gmail.com

mortars has garnered considerable attention to reduce the environmental footprint by cutting the use of cement and/or natural aggregates.

Marble, a type of metamorphic rock that originates from limestone, undergoes an intricate geological journey involving tectonic processes, high pressures, and elevated temperatures near the Earth's crust. These transformative forces result in the neoformation of calcite crystals and the distinctive granoblastic texture characteristic of marble. There are several types of marbles and their purity is determined by the proportions of calcite and dolomite, the primary constituents of marble (Casal Moura et al. 2007). Additionally, marbles can contain a spectrum of minerals like biotite, chlorite, serpentine, amphiboles, hornblende, pyroxenes, garnet, among others. These minerals, in conjunction with the limestone's original attributes, dictate the coloration of the marble, a trait often influenced by its geographical origin (Singh et al. 2017). Marble is frequently referred to as an ornamental stone. It is a durable and abundant stone used as construction material in tiles, lintels, decorative finishing, sculptures, and aggregates. Notably, the marble industry ranks as a pivotal sector in countries' economies endowed with marble reserves (Alyamaç and Aydin 2015).

In the EU-27, mining and quarrying activities are directly responsible for about one-quarter (26.6%) of all waste generated and this share increases even further if downstream activities such as construction are considered (62.5%) (Eurostat 2022). Despite the increased production of this type of stone in Iberian Peninsula during the 21st century, the extraction and processing of ornamental stone encompasses significant environmental impacts and generates large residues (André et al. 2014, DGEG 2022, Kuoribo and Mahmoud 2022, Montani 2020, Rana et al. 2016, Simão et al. 2021). Upcycling ornamental stone wastes (OSW) is critically important for countries like Portugal, where marble exploration has experienced steady growth over the years and where the adoption of circular exploration models for mineral resources needs to be put forward rapidly to comply with the environmental targets laid down by EU 8th Environmental Action Plan (European Comission 2020) and the United Nations' 2030 Agenda and its Sustainable Development Goals (United Nations 2022).

The goal of using marble powder in concrete is related to the environmental impact of cement production. To produce one ton of cement, approximately 1.5 tons of limestone are required. Cement production generates around 0.94 tons of $CO_2$ emissions per ton of cement produced, and certain raw materials, like limestone, have finite reserves that may deplete over time (Rana et al. 2015). Globally, concrete production reached 4.2 billion tons, potentially contributing to 3.95 billion tons of $CO_2$ emissions (Arel 2016). This represents 7%

of greenhouse gas emissions, with predictions suggesting that cement production could rise to approximately 6 billion tons by 2050 (Kore et al. 2020).

Numerous research efforts have been dedicated to exploring the incorporation of marble powder within cementitious matrices, either as an additive alongside traditional components (Mendonça et al. 2021), as a cement replacement (Aliabdo et al. 2014, Bacarji et al. 2013, Lezzerini et al. 2022, Mashaly et al. 2016, Rana et al. 2015, Rodrigues et al. 2015, Singh et al. 2017, Talah et al. 2015) or as a substitute for natural aggregates, like natural sand (Aliabdo et al. 2014, Alyamaç and Aydin 2015, Bourzik et al. 2023, Anitha Selvasofia et al. 2021). For all three alternatives, the authors determined that an optimal substitution ratio falls within the range of 10% to 15% to maintain acceptable mechanical and rheological properties.

Therefore, the addition of this type of waste material in 3D-printed mortars is expected to enhance their environmental performance and may also contribute to improve the printing process and final performance of building components. Expected benefits include: (i) cement and natural aggregates content reduction in the mixture composition of 3D-printed mortars; (ii) a wide palette of colours and textures that can be achieved by the complex chemistry and varied mineralogy of marble powder, improving the commercial attractiveness of 3D-printed mortars; and (iii) opening of new circular business opportunities.

## MATERIALS AND METHODS

### Experimental Plan

In the formulation of 3D-printed mortars with marble powder, a preliminary experimental plan had to be meticulously designed. This plan was carefully structured to encompass the critical aspects associated with the development of these mortars. The ultimate goal was to provide compelling evidence supporting their potential use as architectural components in the future. The process involves a combination of mixing, printing, testing, and analysing by using specialised equipment to ensure accurate and meaningful results.

Several tasks were defined, as seen in Table 1.

**Table 1**   Experimental plan for design and production of 3D-printed mortars with marble powder

| Task | Objective | Methods | Equipment |
|---|---|---|---|
| 1. Mixture design | To assess the feasibility of replacing binder and natural aggregates with marble powder | Formulate different mortar compositions varying proportions of cement, lime, marble powder, sand, superplasticiser, and water | Mixer, scale, cement, lime, marble powder, sand, superplasticizer, and water |
| 2. Fresh state testing | To evaluate fresh state properties of different mortars | *Extrudability*: Measure the capacity of the mortar to be extruded through the 3D printer's nozzle *Buildability*: Assess the mortar's ability to maintain shape and adhere to the previous layer during printing | Piping bag and slump mould |
| 3. Mortar printing | To print mortar samples | Use a 3D printer to create sample specimens from the formulated mixtures | 3D printer |
| 4. Printer calibration | To calibrate the printer's parameters | While printing, define the most appropriate printing speed, extrusion rate and layer height | 3D printer and slicer software |
| 5. Hardened state testing | To assess the hardened state properties | *Mechanical resistance*: Evaluate the compressive and flexural strength of the cured mortars *Capillary water absorption*: Measure the ability of the cured mortars to absorb water | Compression and flexural strength testing equipment, capillary water absorption testing setup |
| 6. Data analysis | To analyse and interpret the test results | Compare the test results against predefined performance criteria and industry standards | Computer |
| 7. Refinement | To make necessary adjustments and refinements | Based on the test results, adjust the mixture proportions and composition as needed for optimal performance | All of the previous equipment |

## 3D PRINTER

A small-size 3D printer *Delta WASP 40100 Clay* was used for the trials, since it is capable of extruding a wide variety of materials, from porcelain, earthenware, stoneware, refractory materials, clays (harder mixtures) to concrete or geopolymers (fluid-dense materials), and thus the ideal option for testing new mixtures and to create edgy design components. The *WASP Manual Feeding Extruder*, for concrete or geopolymer, consists of a cone with a 2.50 l capacity made of stainless-steel that is loaded manually by pouring the material directly into the cone. The flow is controlled by a stepper motor connected to a screw. To keep the proper material rheology and prevent premature hardening, a paddle located inside the cone continuously remixes the material. (attention should be paid here as the paddle is not designed to blend the mixture components; the mixture must be prepared in advance with a suitable mixer). The available stainless-steel nozzles that come with the extruder have a diameter of 6 mm and 8 mm, respectively. Table 2 summarises its printing specifications.

**Table 2**    Printing specifications

| Grain size (mm) | Powder content | Build dimensions (mm) | Extruder/cone capacity (l) | Nozzle diameter (mm) | Layer height (recommended) |
|---|---|---|---|---|---|
| <2 | Very high | Ø 40 × 100 | 2.50 | 6 and 8 | 1–4 |

## Materials Characterisation

For the experimental trials, several materials were used:

- Binders: *White Portland limestone cement CEM II/B-L 32.5R (BR) (EN 197-1:2011 (CEN 2011))* from Secil. It distinguishes itself by being a grey cement with lower $Fe_2O_3$ and alkali content but higher free lime content. To enhance its whiteness, it is ground to a finer consistency than grey Portland cement. This finer grind contributes to faster setting times and higher initial compressive strength. In addition to its unique ability to adapt to the colour of other components, it is widely and readily applied in both the domestic market and a majority of European countries.

  *Hydraulic lime: HL 5 (EN 459-1:2010 (CEN 2010))* from Secil. It is a time-tested material, favoured for historic structures, due to its unique benefits. It facilitates moisture evaporation, safeguarding against potential damage, and offers flexibility to resist cracking. With a moderate setting time, it maintains optimal viscosity for 3D Printing (3DP) while ensuring durability. Notably, it is an eco-friendly choice, producing significantly fewer carbon emissions compared to cement, although it may not match cement's high mechanical strength.

- Aggregate: river sand with grain size <1 mm from MAXMAT.

- Admixture: superplasticizer ViscoCrete-20 HE from Sika. It enables the efficient dispersion of both cement/hydraulic lime and marble powder particles, simultaneously reducing the required water content in the mortar.
- Marble powder was received as waste from furniture manufacturing processes and was used as inert, to partially replace cement and sand. A granulometric analysis of the marble powder became complex and unprofitable due to the heterogeneities of the batches.
- Tap water was used to blend all the components together.

## Mixture Design

The initial challenge of this study was set around designing 3D-printed mixtures using locally sourced raw materials (Portland lime cement, hydraulic lime, marble powder and sand). Given the innovative nature of marble powder and the lack of studies in the existing literature on similar non-standard mixtures, a systematic approach was undertaken to address this challenge, drawing insights from prior research (Hwang and Khoshnevis 2005, Kazemian et al. 2017, Kruger et al. 2019, Lafhaj et al. 2019, Nerella et al. 2016, Paul et al. 2018a, Schröfl et al. 2019, Tay et al. 2016, Teixeira et al. 2021, Ting et al. 2019), and exploiting the authors' previous knowledge in cement-based mixture design for 3DP (Jesus et al. 2023, Teixeira et al. 2021, 2022). Through a methodical and iterative process, the aim was to develop several formulations that seamlessly exhibited key 3DP characteristics (extrudability and buildability) and have a significant reduction in cement and sand content to evaluate the impact of marble powder in enhancing the final element's appearance and properties.

Sand usage should be carefully considered, as it has the potential to block the 8 mm nozzle and can also contribute to a rougher texture in the sample. Then, the best water-to-powder (w/p) or water-to-binder (w/b) ratio could be established. As seen in the equipment calibration, a 3D-printed mortar needs to have pumpability, flowability, extrudability, and buildability to reach printability. Since a pump system was not employed, only extrudability, buildability, and printability were evaluated, as detailed in Table 3. and explained earlier. Mortar properties are intricately linked to the selection and proportion of binders, aggregates, adjuvants, and additives, influencing the fulfilment of tailored properties.

Before testing the material on the printer, the mini-slump test (Gao and Fourie 2015) and a manual extrusion test were performed. Manual extrusion was carried out by using a piping bag with a similar nozzle (in shape and diameter) to the printer. If the mortar can be manually extruded, it will be tested on the printer during the next task. If it does not respect any of the properties presented in Table 3, a reformulation of the composition will be necessary. Also, if the material can be extruded with the piping bag, but not by the printer, it will also have to be redesigned (Figure 1).

**Table 3**   Acceptance criteria for 3D-printed mortars

| 3DP properties | Acceptable | Not acceptable |
|---|---|---|
| Extrudability | Uniform filament | Filament interruptions or width variations |
| Buildability | Stable and uniform layers | Layers with large deformations or collapses |
| Printability | Preservation of extrudability and buildability throughout the entire printing | Lack of extrudability or buildability throughout printing |

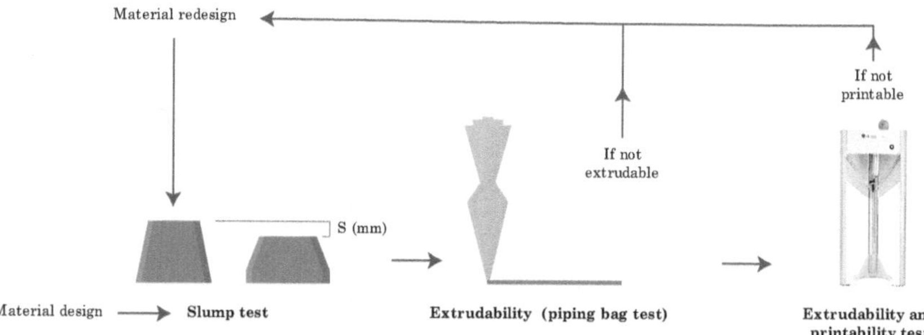


**Figure 1**   Proposed procedure for the development of 3D-printed mortars for *Delta WASP 40100 Clay.*

Tables 4 and 5 gather several compositions that were extrudable through the piping bag, comprising cement, hydraulic lime, marble powder, sand, water and, occasionally, superplasticiser. To evaluate the impact of this residue on the mixture, several laboratory tests were performed, as will be presented below.

Starting with mixture T4, which was successfully used in the printer, the cement content was initially kept constant at 250 g, while sand was gradually replaced with marble powder in increments of 50 g until mix S7 was achieved, which contained no sand.

The second objective was to replace cement with marble powder while keeping the sand content constant at 150 g, resulting in mixture C4. Mixtures S1 and S2 displayed a significant influence of sand on texture, while S5 and S6 had minimal sand content, therefore the sand content of S4 (150 g) was the most appropriate. For type C mixtures, cement was not entirely omitted as these mixtures required a binder. Marble powder was progressively added in 25 g portions to better understand the resulting changes.

**Table 4**  Mixture design of lime-based mortars

| Mortar | Lime (g) | Marble powder (g) | Sand <1 mm (g) | Water (g) | SP (g) | w/p |
|---|---|---|---|---|---|---|
| **L1** | **666.67** | – | **333.33** | **306.67** | – | **0.46** |
| L2 | 600.00 | – | 400.00 | 306.67 | – | 0.51 |
| L3 | 600.00 | – | 400.00 | 300.00 | – | 0.50 |
| **L4** | **550.00** | – | **450.00** | **253.00** | – | **0.46** |
| L5 | 600.00 | – | 400.00 | 280.00 | – | 0.47 |
| **L6** | **600.00** | – | **400.00** | **275.00** | – | **0.46** |
| L7 | 700.00 | – | 300.00 | 322.00 | – | 0.46 |
| L8 | 650.00 | – | 350.00 | 305.00 | – | 0.47 |
| **L9** | **600.00** | – | **400.00** | **185.00** | 3.00 | **0.31** |
| L10 | 700.00 | – | 300.00 | 225.00 | 3.50 | 0.32 |
| L11 | 600.00 | 300.00 | 100.00 | 384.00 | – | 0.43 |
| L12 | 500.00 | 300.00 | 200.00 | 340.00 | – | 0.43 |
| L13 | 500.00 | 400.00 | 100.00 | 394.00 | – | 0.44 |
| L14 | 400.00 | 300.00 | 300.00 | 320.00 | – | 0.46 |
| L15 | 400.00 | 500.00 | 100.00 | 370.00 | – | 0.41 |
| L16 | 400.00 | 600.00 | – | 430.00 | – | 0.43 |
| L17 | 300.00 | 500.00 | 200.00 | 350.00 | – | 0.44 |
| L18 | 200.00 | 600.00 | 200.00 | 400.00 | – | 0.50 |

**Table 5**  Mixture design of cement-based mortars

| Purpose | Mortar | Cement (g) | Marble powder (g) | Sand <1 mm (g) | Water (g) | SP (g) | w/p |
|---|---|---|---|---|---|---|---|
| Printer calibration | T1 | 826.00 | – | 174.00 | 300.00 | – | 0.36 |
| | T2 | 250.00 | 625.00 | 125.00 | 380.00 | – | 0.43 |
| | T3 | 250.00 | 600.00 | 150.00 | 340.00 | – | 0.40 |
| | **T4** | **250.00** | **575.00** | **175.00** | **370.00** | – | **0.45** |
| | **T5** | **250.00** | **575.00** | **175.00** | **250.00** | **2.00** | **0.30** |
| Sand replacement | S1 | 250.00 | 450.00 | 300.00 | 316.00 | – | 0.45 |
| | S2 | 250.00 | 500.00 | 250.00 | 336.00 | – | 0.45 |
| | S3 | 250.00 | 550.00 | 200.00 | 356.00 | – | 0.45 |
| | S4 | 250.00 | 600.00 | 150.00 | 376.00 | – | 0.44 |
| | S5 | 250.00 | 650.00 | 100.00 | 390.00 | – | 0.43 |
| | S6 | 250.00 | 700.00 | 50.00 | 370.00 | – | 0.39 |
| | S7 | 250.00 | 750.00 | – | 420.00 | – | 0.42 |
| Cement replacement | C1 | 325.00 | 525.00 | 150.00 | 312.00 | – | 0.37 |
| | C2 | 300.00 | 550.00 | 150.00 | 316.00 | – | 0.37 |
| | C3 | 275.00 | 575.00 | 150.00 | 320.00 | – | 0.38 |
| | C4 | 250.00 | 600.00 | 150.00 | 324.00 | – | 0.38 |
| | C5 | 225.00 | 625.00 | 150.00 | 330.00 | – | 0.39 |
| | C6 | 200.00 | 650.00 | 150.00 | 340.00 | – | 0.40 |
| | C7 | 175.00 | 675.00 | 150.00 | 350.00 | – | 0.41 |

The methodology for the lime-based mixtures took a different path. Initially, several lime-to-sand ratios were examined, followed by the creation of 8 marble powder variations (L11 to L18). Marble powder quantities ranged from 300 g to 600 g, with corresponding lime quantities ranging from 600 g to 200 g. Simultaneously, the sand component was adjusted from 0 g to 300 g.

Mortars L1 to L10 and T4 were the first to be tested on the printer, while the other ones were successfully extruded through the piping bag. As marked in bold in the above tables, the best results were achieved with L1, L4, L6, L9 and T4 formulations, leading to the conclusion that a w/p ratio close to 0.46 is the most suitable for this specific printer.

## 3D PRINTER CALIBRATION

The printing parameters to be established during the mixture design stage are printing speed, flow rate (extruder rotation speed that controls the amount of extruded material), and nozzle height (Ma et al. 2018, Malaeb et al. 2015).

The printing speed and flow rate should be optimised during the extrudability test. This test consists of printing a long filament; if it is interrupted or there are changes in its width, the printing parameters or the mixture composition will be adjusted. Figure 2(a) shows an example of a mixture with good extrudability, where the printing speed was set at 10 mm/s and the flow rate at 3.30. The walls of the same 3D model were used to assess buildability. Since no deformations or collapses were observed after printing several stacked layers, as shown in the printed sample of Figure 2(b), the mixture was approved. The material passes printability if it keeps extrudability and buildability throughout the entire printing. For these set of calibration testes, mortars T1 to T4 were used.

(a)              (b)

**Figure 2** Printed sample on a round Ø 400 mm wood plate with good:
(a) extrudability; (b) buildability.

Occasionally, the mixture's consistency may be appropriate for printing, but achieving optimal print quality requires fine-tuning of the printing parameters. Figure 3 shows that an adjustment in the printing speed and the flow rate during the printing process (through the printer's control panel) can enhance the extruded filament quality.

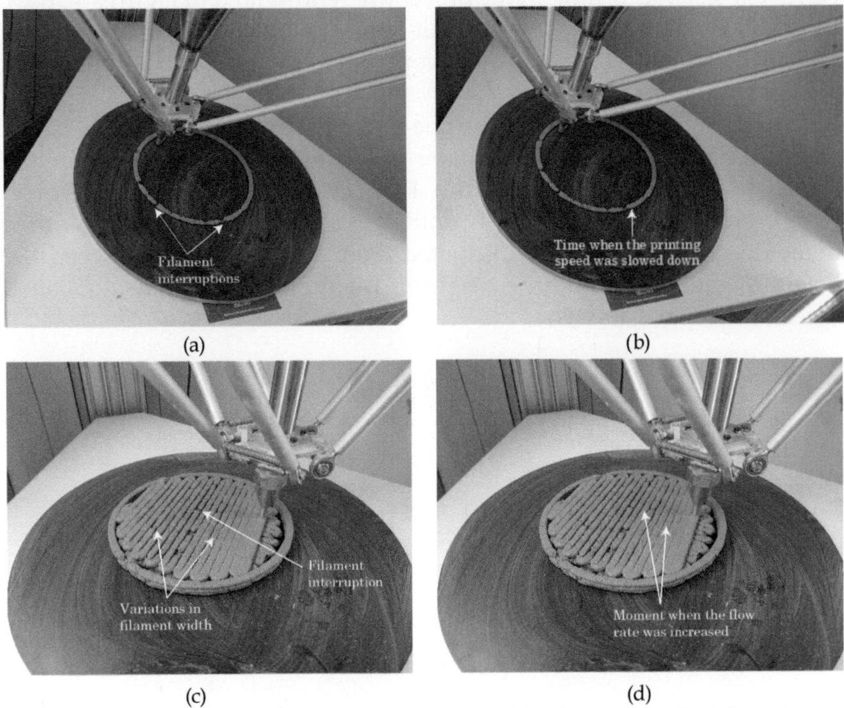


**Figure 3** Calibration of printing parameters: (a) printing speed at 50 mm/s; (b) printing speed at 10 mm/s; (c) flow rate at 150%; (d) flow rate at 330%.

The printing speed and flow rate determined in these printings will be used as a reference for future tests. The layer height was established according to the recommendation of (Bos et al. 2016), where the height of the nozzle must be lower than its nozzle diameter (normally this value corresponds to half the extruder nozzle diameter or even less as seen in (Lin et al. 2018)). In this regard, the height was set to 4 mm, since the nozzle diameter used was 8 mm.

In conclusion, printing speeds of 10 mm/s and a flow rate of 3.30 are acceptable to obtain a proper extrudability and buildability for mortars without admixtures/additives and for this printer, after trial-error attempts.

# FRESH AND HARDENED STATE TESTS

## Fresh State Tests

After mixing, the fresh mortar was evaluated through slump and manual extrusion tests, according to the procedure represented in Figures 4 and 5, respectively. These tests, carried out before the printing task, cover the main 3DP properties for mortars: workability, extrudability, buildability, and, finally, printability.

The slump test (similar to the one proposed by (Gao and Fourie 2015) was performed to assess the shape retention capacity of the material. To do so, mortar was cast into a PVC cylindrical mould (60 mm Ø by 60 mm in height (Figure 4) until it reached the top and was then demoulded. The slump is calculated from the difference between the height of the mould and the height of the mortar. The values obtained are indicated in Table 6.

**Figure 4**   Slump test probe and mould.

**Table 6**   Slump test values

| Mortar | Slump (mm) | Mortar | Slump (mm) | Mortar | Slump (mm) |
|--------|-----------|--------|-----------|--------|-----------|
| S1 | 8 | C1 | 3 | L3 | 6 |
| S2 | 6 | C2 | 5 | L11 | 4 |
| S3 | 6 | C3 | 4 | L12 | 3 |
| S4 | 7 | C4 | 3 | L13 | 5 |
| S5 | 6 | C5 | 2 | L14 | 6 |
| S6 | 3 | C6 | 4 | L15 | 3 |
| S7 | 5 | C7 | 4 | L16 | 4 |
|  |  |  |  | L17 | 4 |
|  |  |  |  | L18 | 8 |

The extrudability test (Figure 5(a)) evaluates whether the material exhibits the required workability to smoothly pass through the nozzle, without encountering issues like clogging or excessive force. The goal was to produce a continuous and uniform filament without any defects, using a piping bag. The initial aim of this test was to create a 20 cm filament. If this was successfully verified, the extrusion process would be repeated with 4 more filaments, with 2 minutes intervals between each.

If the material proved to be extrudable, the buildability test would be conducted right after (Figure 5(b)). It evaluates the material's capacity to remain stable as additional layers are deposited above it, while also verifying the adhesion between these layers. The objective of this test was to successfully stack 3 to 5 layers of 20 cm filaments on top of one another. The assessment was made visually, according to the criteria defined in Table 3.

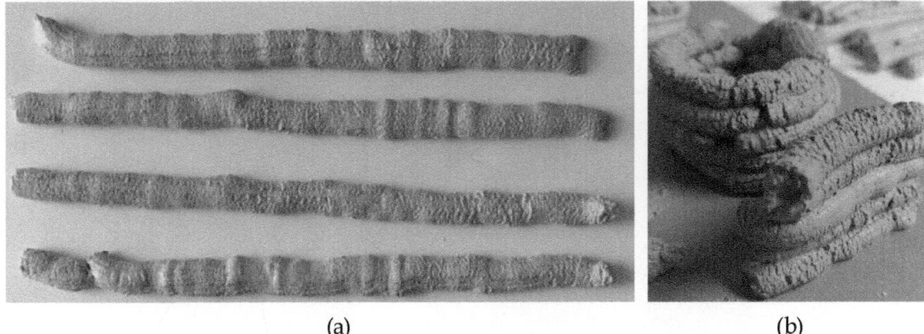

(a)          (b)

**Figure 5** Fresh state tests: (a) manual extrudability test;
(b) manual buildability test.

## Hardened State Tests

Later, some mixtures were moulded to assess their flexural and compressive strength, both expressed in MPa. Flexural strength serves as an indirect indicator of a material's tensile strength, reflecting its capacity to withstand deformation under a load. Typically, this property is determined through a 3-point bending test, where the sample is supported at both ends while a specific load is applied at its midpoint (Figure 6(a)). The flexural strength $R_f$ (MPa) is determined by Equation (1), where $b$ is the size (mm) of the square section of the prism, $F_f$ is the load (N) applied to the middle of the prism at fracture and $l$ (mm) is the distance between the supports.

$$R_f \frac{1.5 \times F_f \times l}{b^3} \tag{1}$$

The test concludes when the sample eventually fractures into two parts. These two parts are latter submitted to a compressive strength test that evaluates the material's resistance to be compressed or pushed together. This property is quantified by dividing the applied load at the point of failure by the cross-sectional area of the gauge section (Figure 6(b)), as defined in Equation (2):

$$R_c \frac{F_c}{1600} \tag{2}$$

where $R_c$ is the compressive strength (MPa), $F_c$ is the maximum load at fracture (N) and 1600 is the area of the platens (40 mm × 40 mm). These tests followed, in part, the procedure suggested by EN 196-1:2016 (CEN 2016). For each mixture, three prismatic specimens with dimensions of 40 mm × 40 mm × 160 mm were produced. Cement-based and lime-based mortars were demoulded after 1 and 5 days, respectively. Then, they were stored in a cure room with a controlled temperature of 20 ± 2°C and 50 ± 2% relative humidity (RH) and removed at the age of 28 days to be subjected to the abovementioned mechanical strength tests.

(a)                                          (b)

**Figure 6**   Mechanical strength tests: (a) three points flexural strength test; (b) compressive strength test.

The water absorption per capillarity represents the amount of water absorbed per unity area due to capillary suction forces (Figure 7). This parameter, represented as $A_w$ and expressed in kg/(m²$\sqrt{s}$), was determined according to ISO 15148:2002 (ISO 2002) guidelines. The test involves immersing a material specimen's lower surface in water, with the water level kept 5 ± 2 mm above the specimen's highest point. The mass variation of each specimen was recorded over time (after 5 min, 10 min, 12 min, 20 min, 30 min, 40 min, 1 h, 1.5 h, 2 h, 4 h, 6 h, 8 h and 24 h). To prevent water intrusion from the sides and guarantee unidirectional flux, the specimen edges were previously sealed with an epoxy resin. The specimens were conditioned at 23 ± 2°C and 50 ± 5% RH (laboratory environment) until their mass became constant. Subsequently, their initial masses were accurately measured. The mass variation, $\Delta m_t$ (kg/m²), is computed using Equation (3), where $A$ denotes the specimen's face area in square meters (m²), $m_i$ is the initial mass (kg), and $m_t$ is the mass at a specific time (kg) $t$. The water absorption coefficient is expressed in Equation (4), where $\Delta m_{tf}$ is the result mass variation at the end of the test (24 h).

$$\Delta m_t \frac{m_t - m_i}{A} \tag{3}$$

The water absorption coefficient is expressed in Equation (4), where $\Delta m_{tf}$ is the result mass variation at the end of the test (24 h).

$$A_{w,24} = \frac{\Delta m_{tf}}{\sqrt{86400}} \tag{4}$$

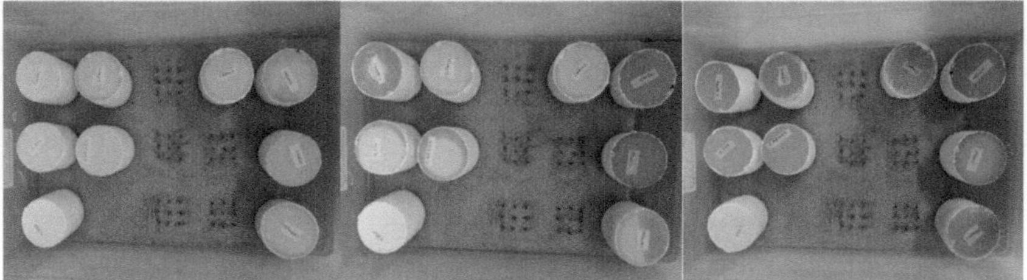

**Figure 7**   Setup for water absorption test.

## RESULTS AND DISCUSSION

### Mixture Design

During the mixture design stage, the printing speed was kept at 100% in the printer's control panel (10 mm/s, as defined in *Simplify 3D* software) and the extrusion flow defined in *Simplify 3D* was 3.30, changed during the printing process whenever needed on the printer's screen.

Mixture L1 was the first formulation to exhibit favourable attributes essential for 3DP, including extrudability and buildability. Subsequently, while keeping a constant solid content of 1 kg, there was a reduction in the binder component, accompanied by a proportional increase in the aggregate content. This adjustment served as an experimental exploration to determine the aggregate limit at which the printer effectively works, providing valuable insights for the formulation of future mixtures. Furthermore, it was observed that the water content required for subsequent mixtures could be previously defined based on the w/p ratio of the first mixture, allowing for some refinements as necessary.

From Tables 4 and Table 5, the next mortars to be tested in the printer were L3, L14, L18, S2, S3, S4, S7 and C7, since they had the closest w/p values to 0.46.

Despite being extrudable to a certain extent, mixtures L3 and L7 suffered elastic buckling (Figure 8), a structural instability mechanism characterised by a loss of geometric stability, often leading to gradual lateral deformations of the printed element's walls (Suiker et al. 2020). In this scenario, it was observed that the low aggregate-to-binder (a/b) ratio of mixture L7 (0.43) was a major factor in the occurrence of this phenomenon (despite having a w/p ratio of 0.50). The study conducted by Sasikumar et al. (Sasikumar et al. 2023) proved that a significant coarse aggregate ratio (of about 40%) can improve buildability. It also seems that excess water can dilute the concentration of cementitious materials, slowing down the hydration process and promoting longer curing periods, affecting the material's capacity to support the weight of subsequent printed layers. As a result, the lower layers might deform or deflect under the load, leading to instability and buckling in the structure.

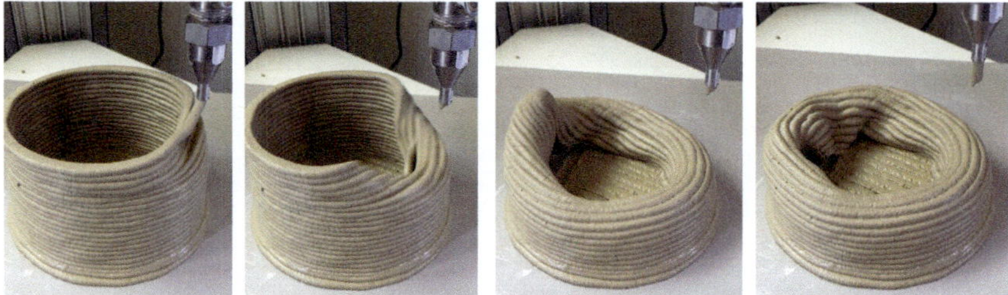

**Figure 8**     Elastic buckling in printed sample of mixture L7.

Another problem associated with excess water is the eventual plastic shrinkage cracking, a prevalent challenge in freshly placed concrete, resulting from the evaporation of pore water and the development of tensile strains exceeding the mortar's capacity (Combrinck and Boshoff 2019, Ravina and Shalon 1968). This issue is amplified in 3D-printed cementitious mortars, which, after printing, are more exposed to environmental factors such as temperature, humidity, and wind due to the absence of conventional formwork, accelerating hydration reactions (Almusallam 2001, Holt and Leivo 2004a, Ortiz et al. 2005). The low aggregate content and absence of coarse aggregates in 3D-printed mortars contribute to increased shrinkage in its plastic and hardened states (Almusallam et al. 1998, Paul et al. 2018b, Shaeles and Hover 1988). Additionally, a lower water-to-cement ratio (w/c) can significantly enhance autogenous shrinkage (Holt and Leivo 2004b). The influence of superplasticizers on this issue varies in literature (Combrinck et al. 2019, Sayahi et al. 2019, Turcry and Loukili 2006), making careful material selection and mix proportioning vital for managing plastic shrinkage cracking in 3D-printed mortars while ensuring pumpability and buildability. Mixture L3, that has a w/p ratio of 0.46, suffered from both elastic buckling and plastic shrinkage cracking (Figure 9), supporting the previous information from the literature.

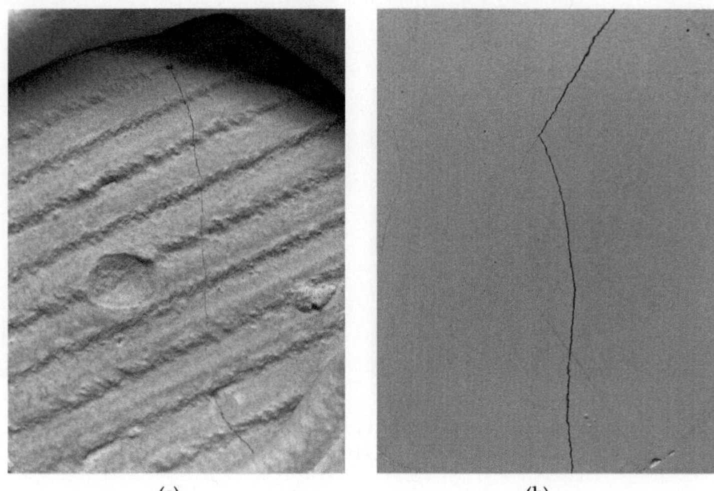

(a)                                        (b)

**Figure 9**  Plastic shrinkage cracking in: (a) top of the base;
(b) bottom of the base.

For this printer, the use of superplasticiser (0.5% of binder mass) and the reduction of the layer height from 4 mm to 2.50 mm were crucial to tackle these phenomena, as was increasing the mechanical resistance of the mixture and improving the workability, while maintaining the desired fluidity, as described in the study of (Dias et al. 2023). This is explained by the water reduction of more than 30 per cent. It also allowed for a substantial boost in printing speed, rising from 10 mm/s to 35 mm/s (a 350% increase), significantly reducing overall printing time. These additives promote superior strength development, reduce segregation and bleeding tendencies, while optimising printing parameters. They also expand design possibilities by enabling the incorporation of various additives and enhancing the overall quality, durability, and precision of 3D-printed structures. As portrayed in Figure 10. during the extrusion of mixture T4, with no superplasticiser, the printing parameters were optimised to overcome the "cone effect", resulting in an irregular surface due to the variation in the layers' width. This effect is characterised by the tendency of layers near the base of a printed object to be slightly wider than those positioned higher in the structure. It is primarily attributed to the extended cooling and solidification time of the lower layers before the subsequent layer is added on top. It can often contribute to elastic buckling, as described above. Contrarily, the printing parameters were consistent from the beginning to the end of printing with mixture T5 (based on T4), resulting in a uniform surface and a shinier finish.

(a)                                             (b)

**Figure 10**   Printed samples with proper printability: (a) mixture T4;
(b) mixture T5.

Since this study involves the use of marble powder in mixtures for 3DP, it was important to evaluate its water necessities for exterior walls building elements. The graph of Figure 11 describes this correlation.

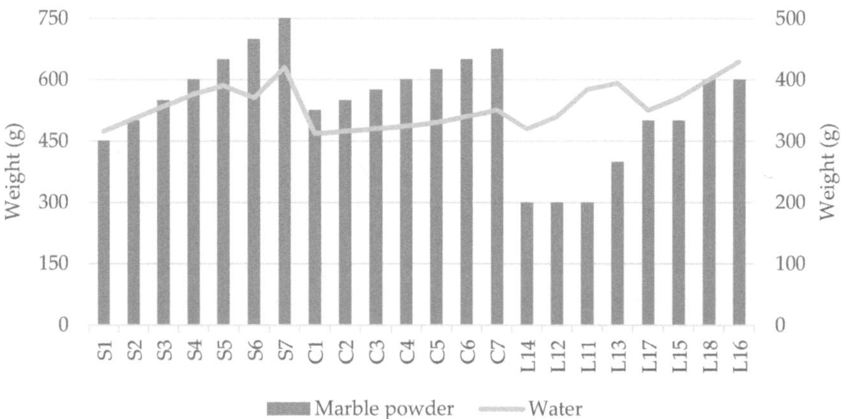


**Figure 11**   Water content in 3D-printed mixtures as function of marble powder.

Marble powder has a substantial impact on the water requirement in all types of mixtures. When comparing mixtures with similar amounts of marble powder, type C mixtures demand less water than type S mixtures. Type C mixtures generally contain less sand compared to type S mixtures, with more binder and aggregates, and thus offer an explanation for this trend. Interestingly, even though type L mixtures contain less marble powder than the others and have a low fine aggregate content, their water consumption remains quite similar.

## Fresh State Tests

Extrudability and buildability were assessed through customised tests. As stated above, all the mixtures were extrudable through the piping bag and showed proper shape retention.

Figure 12 presents a correlation graph of slump values and the w/p ratio of the tested mixtures.

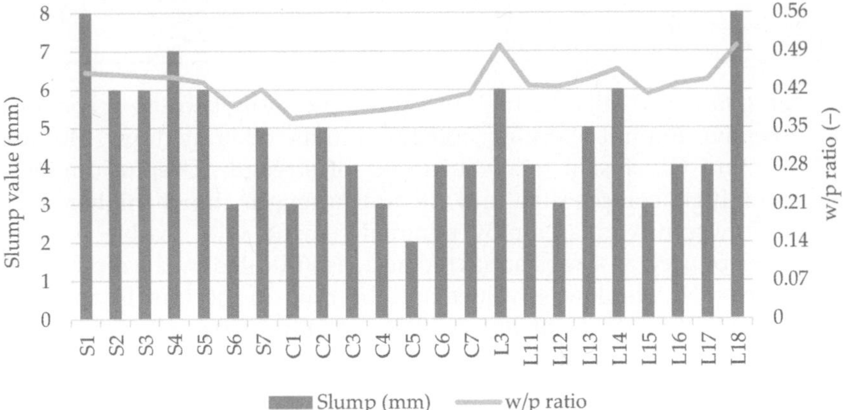


**Figure 12**   Slump test values and mortars' w/p ratio.

Certain mixtures demonstrate that an increase in the w/p ratio corresponds to higher slump values, signifying a greater fluidity in the mixture.

As previously noted, the recommended w/p ratio for the printer stands at approximately 0.46, which, according to the graph, corresponds to a slump of 6 mm. Remarkably, most of the mixtures earmarked for printing (L3, L14, L18, S2, S3, S4, S7, and C7) align with this specific slump value of 6 mm. This convergence emphasizes that, beyond the used ratio, the slump test serves as a reliable indicator of the mixtures' printability.

## Hardened State Tests

Since there were only six moulds available, each containing three slots, it was possible to select only six mixtures from all those tested in the manual extrusion test. Therefore, the most extreme cases were considered: T4, which was tested in the printer; S7, lacking sand; C7, containing more marble powder than T4; T1, with significantly more cement than the others; L3, the base mixture with lime (without marble powder); and L16, which consisted solely of marble powder (without sand).

Average values of flexural and compressive strength and their respective standard deviation (S.D.) are presented in Table 7 where mp/b is the powder-to-binder ratio.

**Table 7**   Hardened state tests results

| Mortar | Flexural strength (MPa) | | Compressive strength (MPa) | | mp/b |
| --- | --- | --- | --- | --- | --- |
| | Mean | S.D. | Mean | S.D. | |
| T4 | 1.76 | 0.13 | 3.58 | 0.31 | 2.30 |
| C7 | 1.43 | 0.06 | 4.47 | 0.17 | 3.86 |
| S7 | 2.12 | 0.20 | 5.84 | 0.74 | 3.00 |
| L3 | 1.85 | 0.71 | 6.16 | 0.33 | 0.00 |
| L16 | 0.77 | 0.19 | 2.07 | 0.05 | 1.50 |
| T1 | 7.90 | 0.98 | 58.32 | 3.76 | 0.00 |

T1 sample exhibits the highest values, primarily because it contains an excessive amount of cement compared to the other mixtures. Mortars L3, S7, and C7 closely follow in terms of strength. Despite cement's inherent high compressive and flexural strength (much higher than lime's), these mixtures employ minimal cement, with marble powder being introduced as a substantial replacer, as indicated by the mp/b ratio.

Conversely, the lowest values are observed in mortars T4 and L16. These mixtures stand out due to their notably high mp/b ratio when compared to the others. The excessive use of water in these mixtures could also contribute to this outcome.

Water absorption by partial immersion results at 24 h are coupled in Table 8 and the mass change over time ($\Delta m_t$) graphic is represented in Figure 13.

**Table 8**   Water absorption by partial immersion results

| Mortar | Initial mass, $m_i$ (kg) | Area, $A$ (m$^2$) | Mass variation, $\Delta m_{t,24}$ (kg/m$^2$) | Water absorption coefficient, $A_{w,24}$ (kg/(m$^2$√s)) |
| --- | --- | --- | --- | --- |
| T1 | 0.3286 | | 9.9647 | 0.0339 |
| S7 | 0.2295 | | 23.9929 | 0.0816 |
| C7 | 0.2409 | 0.00283 | 23.4982 | 0.0799 |
| T4 | 0.2511 | | 21.6961 | 0.0738 |
| L3 | 0.2832 | | 15.6184 | 0.0531 |
| L16 | 0.2127 | | 23.0742 | 0.0785 |

As previously explained, the water absorption coefficient ($A_{w,24}$) of the specimens varies in relation to their mass change ($\Delta m_t$). Therefore, a notable increase in mass indicates a higher water absorption rate by the material.

The values recorded in the table significantly exceed those documented in literature (Pessoa et al. 2023).

Mortars S7 and C7 demonstrated the weakest performance and simultaneously had the highest mp/b ratios. Following closely behind are L16 and T4, which rank 3rd and 4th in terms of marble powder concentration. This suggests that the residue indeed exhibits a substantial water-absorbing capacity and possibly possesses a highly porous structure. Lastly, mortars L3 and T1 stand apart from the rest as they do not incorporate marble powder in their composition. Their distinct behaviour underscores the influence of marble powder on water absorption characteristics within these mortar matrices.

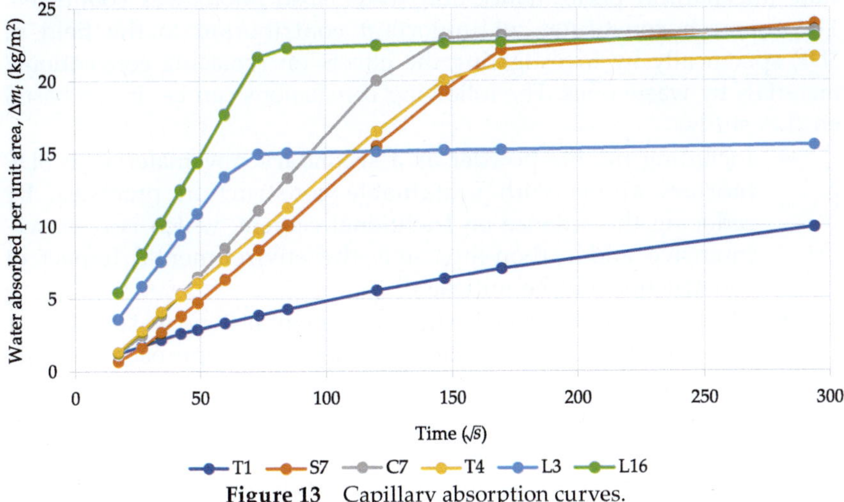


**Figure 13**   Capillary absorption curves.

Furthermore, as depicted in Figure 13, it becomes evident that lime-based mixtures exhibit relatively low water absorption after one hour and thirty minutes from the beginning of the test. This is reflected in the minimal change in mass. In contrast, the cement mixtures demonstrate an increasing water absorption rate beyond this point, indicated by the steeper slope of the curves compared to earlier stages.

This study underscores the importance of carefully considering marble powder content when designing mortars, especially with regard to their intended application and desired porosity. It also highlights the potential of marble powder as a key factor contributing to the porosity of construction materials, thereby influencing their suitability for various construction and engineering purposes. Since these compositions are designed for latter development of decorative products, where mechanical and hygric properties are not a critical factor, it can be concluded that the achieved results are satisfactory and will not comprise the initial purpose of the study. However, further research into optimising marble powder incorporation is warranted to harness its benefits while managing its impact on water absorption, porosity and mechanical properties.

## CONCLUSIONS

In this study, several mixtures for 3DP were designed and tested, where Portland lime cement, hydraulic lime and sand were replaced by marble powder. A further characterisation of some mortars in the hardened state (mechanical performance and water absorption) was conducted. The research constitutes an important contribution to the field of 3DP, specifically by investigating the effects on replacing conventional materials by waste ones. The following conclusions can be drawn based on this study:

- Including marble powder as a secondary raw material in 3DP mortars aligns with sustainable construction practices. By reducing the reliance on traditional cement, which is resource-intensive and carbon-intensive, the environmental impact of construction can be mitigated;
- The use of marble powder from ornamental stone waste presents an effective waste valorisation strategy. This approach not only reduces the disposal of OSW but also transforms it into a valuable resource – increasing the materials palette for 3DCP, contributing to circular economy principles;
- Depending on its granulometry, marble powder introduces a diverse palette of colours and textures to 3D-printed mortars, enhancing their aesthetic appeal. This versatility can expand the design possibilities for architectural components and decorative features;
- Despite maintaining or even enhancing key properties such as workability and extrudability, marble powder is not capable of optimising the mechanical properties of 3D-printed mortars;
- The use of superplasticiser was crucial to avoid elastic buckling and plastic shrinkage cracking by substantially reducing the water content of mixtures (in more than 30%);
- The printed components attained a curiously soft and shiny finish, attributed not only to the diminished aggregate content but also to the incorporation of the powder and the inclusion of a superplasticiser.

Overall, this work underscores the feasibility and benefits of utilising marble powder in 3D-printed mortars, contributing to sustainable practices, waste reduction, and developing environmentally friendly building materials. As the Construction industry continues to embrace sustainability, the integration of marble powder in 3D-printed mortars represents a meaningful step towards more eco-conscious and aesthetically diverse construction solutions; however, further optimisation and careful characterisation of the material is still necessary.

## ACKNOWLEDGEMENTS

Manuel Jesus would like to thank FCT for financial support through the doctoral grant UI/BD/151533/2021. João Teixeira would like to thank FCT – Fundação para a Ciência e a Tecnologia, I.P. for the PhD grant 2020.07482.BD through FSE/NORTE 2020 funding (https://doi.org/10.54499/2020.07482.BD). Bárbara Rangel would like to thank Base Funding – UIDB/00145/2020 of the CEAU – Center for Studies in Architecture and Urbanism.

Finally, this work was financially supported by: Base Funding – UIDB/04708/2020 with DOI 10.54499/UIDB/04708/2020 (https://doi.org/10.54499/UIDB/04708/2020) and Programmatic Funding – UIDP/04708/2020 with DOI 10.54499/UIDP/04708/2020 (https://doi.org/10.54499/UIDP/04708/2020) of the CONSTRUCT – Instituto de I&D em Estruturas e Construções – funded by national funds through the FCT/MCTES (PIDDAC).

## REFERENCES

Aliabdo, A.A., Abd Elmoaty, A.E.M. and Auda, E.M. 2014. Re-use of waste marble dust in the production of cement and concrete. Constr. Build. Mater. 50: 28–41. https://doi.org/10.1016/j.conbuildmat.2013.09.005.

Almusallam, A.A., Maslehuddin, M., Abdul-Waris, M. and Khan, M.M. 1998. Effect of mix proportions on plastic shrinkage cracking of concrete in hot environments. Constr. Build. Mater. 12(6–7): 353–358. https://doi.org/10.1016/S0950-0618(98)00019-1.

Almusallam, A.A. 2001. Effect of environmental conditions on the properties of fresh and hardened concrete. Cem. Concr. Compos. 23(4–5): 353–361. https://doi.org/10.1016/S0958-9465(01)00007-5.

Alyamaç, K.E. and Aydin, A.B. 2015. Concrete properties containing fine aggregate marble powder. KSCE J. Civ. Eng. 19(7): 2208–2216. https://doi.org/10.1007/s12205-015-0327-y.

André, A., de Brito, J., Rosa, A. and Pedro, D. 2014. Durability performance of concrete incorporating coarse aggregates from marble industry waste. J. Cleaner Prod. 65: 389–396. https://doi.org/10.1016/j.jclepro.2013.09.037.

Anitha Selvasofia, S.D., Dinesh, A., and Sarath Babu, V. 2021. Investigation of waste marble powder in the development of sustainable concrete. Mater. Today Proc. 44: 4223–4226. https://doi.org/10.1016/j.matpr.2020.10.536.

Arel, H.S. 2016. Recyclability of waste marble in concrete production. J. Cleaner Prod. 131: 179–188. https://doi.org/10.1016/J.JCLEPRO.2016.05.052.

Bacarji, E., Toledo Filho, R.D., Koenders, E.A.B., Figueiredo, E.P. and Lopes, J.L.M.P. 2013. Sustainability perspective of marble and granite residues as concrete fillers. Constr. Build. Mater. 45: 1–10. https://doi.org/10.1016/j.conbuildmat.2013.03.032.

Bos, F., Wolfs, R., Ahmed, Z. and Salet, T. 2016. Additive manufacturing of concrete in construction: potentials and challenges of 3D concrete printing. Virtual Phys. Prototyping. 11(3): 209–225. https://doi.org/10.1080/17452759.2016.1209867.

Bourzik, O., Baba, K., Akkouri, N. and Nounah, A. 2023. Effect of waste marble powder on the properties of concrete. Mater. Today Proc. 72: 3265–3269. https://doi.org/10.1016/J.MATPR.2022.07.184.

Casal Moura, A., Carvalho, C., Almeida, I.A., Saúde, J.G., Farinha Ramos, J.M., Augusto, J.P., et al. 2007. Mármores e calcários ornamentais de Portugal (A. Casal Moura, Ed.). INETI - Instituto Nacional de Engenharia, Tecnologia e Inovação, I.P.

CEN. 2010. EN 459-1:2010 - Building lime - Part 1: Definitions, specifications and conformity criteria.

CEN. 2011. EN 197-1:2011 - Cement - Part 1: Composition, specifications and conformity criteria for common cements.

CEN. 2016. EN 196-1:2016 - Methods of testing cement - Part 1: Determination of strength.

Combrinck, R. and Boshoff, W.P. 2019. Tensile properties of plastic concrete and the influence of temperature and cyclic loading. Cem. Concr. Compos. 97: 300–311. https://doi.org/10.1016/j.cemconcomp.2019.01.002.

Combrinck, R., Kayondo, M., le Roux, B.D., de Villiers, W.I. and Boshoff, W.P. 2019. Effect of various liquid admixtures on cracking of plastic concrete. Constr. Build. Mater. 202: 139–153. https://doi.org/10.1016/J.CONBUILDMAT.2018.12.060.

DGEG. 2022. Dados Globais da Indústria Extractiva - Produção. https://www.dgeg.gov.pt/pt/estatistica/geologia/dados-globais-da-industria-extractiva/producao/

Dias, B.D., Rocha, D., Faria, P., Lucas, S.S., Silva, V.A., Lobo, B., et al. 2023. Limes with hydraulic properties for 3D printing mortars. pp. 41–50. *In*: Gaspar, F. and Mateus, A. (eds). Sustainable and Digital Building. Springer International Publishing. https://doi.org/10. 1007/978-3-031-25795-7_3.

European Comission. 2020. Decision of the European Parliament and of the Council on a General Union Environment Action Programme to 2030. https://ec.europa.eu/environment/pdf/8EAP/2020/10/8EAP-draft.pdf

Eurostat. 2022. Generation of waste by waste category. Http://Appsso.Eurostat.Ec.Europa.Eu/Nui/SubmitViewTableAction.Do.

Gao, J. and Fourie, A. 2015. Spread is better: An investigation of the mini-slump test. Miner. Eng. 71: 120–132. https://doi.org/10.1016/j.mineng.2014.11.001.

Holt, E. and Leivo, M. 2004. Cracking risks associated with early age shrinkage. Cem. Concr. Compos. 26(5): 521–530. https://doi.org/10.1016/S0958-9465(03)00068-4.

Hwang, D. and Khoshnevis, B. 2005. An innovative construction process-contour crafting (CC). Proceedings of the 22nd International Symposium on Automation and Robotics in Construction. International Association for Automation and Robotics in Construction (IAARC), Ferrara, Italy. https://doi.org/10.22260/ISARC2005/0004.

ISO. 2002. ISO 15148:2002 - Hygrothermal performance of building materials and products - Determination of water absorption coefficient by partial immersion.

Jesus, M., Teixeira, J., Alves, J.L., Pessoa, S., Guimarães, A.S. and Rangel, B. 2023. Potential use of sugarcane bagasse ash in cementitious mortars for 3D printing. pp. 89–103. *In*: da Silva, L.F.M. (ed.). Materials Design and Applications IV. Springer International Publishing. https://doi.org/10.1007/978-3-031-18130-6_7.

Kazemian, A., Yuan, X., Cochran, E. and Khoshnevis, B. 2017. Cementitious materials for construction-scale 3D printing: Laboratory testing of fresh printing mixture. Constr. Build. Mater. 145: 639–647. https://doi.org/10.1016/j.conbuildmat.2017.04.015.

Kore, S.D., Vyas, A.K. and Syed Ahmed Kabeer, K.I. 2020. A brief review on sustainable utilisation of marble waste in concrete. Int. J. Sustainable Eng. 13(4): 264–279. https://doi.org/10.1080/19397038.2019.1703151.

Kruger, J., Zeranka, S. and van Zijl, G. 2019. 3D concrete printing: A lower bound analytical model for buildability performance quantification. Autom. Constr. 106: 102904. https://doi.org/10.1016/j.autcon.2019.102904.

Kuoribo, E. and Mahmoud, H. 2022. Utilisation of waste marble dust in concrete production: A scientometric review and future research directions. J. Cleaner Prod. 374: 133872. https://doi.org/10.1016/j.jclepro.2022.133872.

Lafhaj, Z., Rabenantoandro, A.Z., el Moussaoui, S., Dakhli, Z. and Youssef, N. 2019. Experimental approach for printability assessment: Toward a practical decision-making framework of printability for cementitious materials. Buildings. 9(12): 245. https://doi.org/10.3390/buildings9120245.

Lezzerini, M., Luti, L., Aquino, A., Gallello, G. and Pagnotta, S. 2022. Effect of marble waste powder as a binder replacement on the mechanical resistance of cement mortars. Appl. Sci. 12(9): 4481. https://doi.org/10.3390/app12094481.

Lin, J.C., Wang, J., Wu, X., Yang, W., Zhao, R.X. and Bao, M. 2018. Effect of Processing Parameters on 3D Printing of Cement - based Materials. E3S Web Conf. 38: 03008. https://doi.org/10.1051/e3sconf/20183803008.

Ma, G., Li, Z. and Wang, L. 2018. Printable properties of cementitious material containing copper tailings for extrusion based 3D printing. Constr. Build. Mater. 162: 613–627. https://doi.org/10.1016/j.conbuildmat.2017.12.051.

Malaeb, Z., Hachem, H., Tourbah, A., Maalouf, T., Zarwi, N. and Hamzeh, F. 2015. 3D concrete printing: Machine and mix design. IJCIET. 6: 14–22. https://www.researchgate.net/publication/280488795_3D_Concrete_Printing_Machine_and_Mix_Design.

Mashaly, A.O., El-Kaliouby, B.A., Shalaby, B.N., El-Gohary, A.M. and Rashwan, M.A. 2016. Effects of marble sludge incorporation on the properties of cement composites and concrete paving blocks. J. Cleaner Prod. 112: 731–741. https://doi.org/10.1016/j.jclepro.2015.07.023.

Mendonça, A.M.G.D., Duarte, E.V. de N., Lira, Y.C., Silva, C.C.V.P. da, Luz, T.E.B., Guerra, T.D., et al. 2021. Utilização do resíduo de mármore na produção de argamassa. Brazilian J. Dev. 7(5): 44238–44247. https://doi.org/10.34117/bjdv.v7i5.29212.

Montani, C. 2020. XXXI Report marble and stones in the world. https://issuu. com/marmonews/docs/impaginato_xxxi_mailing.

Nerella, V.N., Krause, M., Näther, M. and Mechtcherine, V. 2016. Studying Printability of Fresh Concrete for Formwork Free Concrete On-site 3D Printing Technology (CONPrint3D). https://www.researchgate.net/publication/2968 17129_Studying_printability_of_fresh_concrete_for_formwork_free_ Concrete_on-site_3D_Printing_technology_CONPrint3D.

Ortiz, J., Aguado, A., Agulló, L. and García, T. 2005. Influence of environmental temperatures on the concrete compressive strength: Simulation of hot and cold weather conditions. Cem. Concr. Res. 35(10): 1970–1979. https://doi.org/ 10.1016/J.CEMCONRES.2005.01.004.

Paul, S.C., Tay, Y.W.D., Panda, B. and Tan, M.J. 2018a. Fresh and hardened properties of 3D printable cementitious materials for building and construction. Arch. Civ. Mech. Eng. 18(1): 311–319. https://doi.org/10.1016/j. acme.2017.02.008.

Paul, S., van Zijl, G., Tan, M. J. and Gibson, I. 2018b. A review of 3D concrete printing systems and materials properties: Current status and future research prospects. Rapid Prototyping J. 24(4): 784–798. https://doi.org/10.1108/RPJ-09-2016-0154.

Pessoa, S., Jesus, M., Guimarães, A.S., Lucas, S.S. and Simões, N. 2023. Experimental characterisation of hygrothermal properties of a 3D printed cementitious mortar. Case Stud. Constr. Mater. 19: e02355. https://doi.org/10. 1016/j.cscm.2023.e02355.

Rana, A., Kalla, P. and Csetenyi, L.J. 2015. Sustainable use of marble slurry in concrete. J. Cleaner Prod. 94: 304–311. https://doi.org/10.1016/J.JCLEPRO.2015. 01.053.

Rana, A., Kalla, P., Verma, H.K. and Mohnot, J.K. 2016. Recycling of dimensional stone waste in concrete: A review. J. Cleaner Prod. 135: 312–331. https://doi. org/https://doi.org/10.1016/j.jclepro.2016.06.126.

Ravina, D. and Shalon, R. 1968. Plastic Shrinkage Cracking. ACI Journal Proceedings, 65(4). https://doi.org/10.14359/7473.

Rodrigues, R., de Brito, J. and Sardinha, M. 2015. Mechanical properties of structural concrete containing very fine aggregates from marble cutting sludge. Constr. Build. Mater. 77: 349–356. https://doi.org/10.1016/j.conbuildm at.2014.12.104.

Sasikumar, A., Balasubramanian, D., Senthil Kumaran, M.S. and Govindaraj, V. 2023. Effect of coarse aggregate content on the rheological and buildability properties of 3D printable concrete. Constr. Build. Mater. 392: 131859. https:// doi.org/10.1016/J.CONBUILDMAT.2023.131859.

Sayahi, F., Emborg, M., Hedlund, H. and Cwirzen, A. 2019. Plastic shrinkage cracking of self-compacting concrete: Influence of capillary pressure and dormant period. Nordic Concrete Research. 60(1): 67–88. https://doi.org/10. 2478/ncr-2019-0012.

Schröfl, C., Nerella, V.N. and Mechtcherine, V. 2019. Capillary water intake by 3D-Printed concrete visualised and quantified by neutron radiography. pp. 217–224. *In*: Wangler, T. and Flatt, R.J. (eds). First RILEM International

Conference on Concrete and Digital Fabrication – Digital Concrete 2018. Springer International Publishing. https://doi.org/10.1007/978-3-319-99519-9_20.

Shaeles, C.A. and Hover, K.C. 1988. Influence of mix proportions and construction operations on plastic shrinkage cracking in thin slabs. ACI Mater. J. 85(6): 495–504.

Simão, L., Souza, M.T., Ribeiro, M.J., Klegues Montedo, O.R., Hotza, D., Novais, R.M., et al. 2021. Assessment of the recycling potential of stone processing plant wastes based on physicochemical features and market opportunities. J. Cleaner Prod. 319: 128678. https://doi.org/10.1016/j.jclepro.2021.128678.

Singh, M., Choudhary, K., Srivastava, A., Singh Sangwan, K. and Bhunia, D. 2017. A study on environmental and economic impacts of using waste marble powder in concrete. J. Build. Eng. 13: 87–95. https://doi.org/10.1016/j.jobe.2017.07.009.

Suiker, A.S.J., Wolfs, R.J.M., Lucas, S.M. and Salet, T.A.M. 2020. Elastic buckling and plastic collapse during 3D concrete printing. Cem. Concr. Res. 135: 106016. https://doi.org/10.1016/j.cemconres.2020.106016.

Talah, A., Kharchi, F. and Chaid, R. 2015. Influence of marble powder on high performance concrete behavior. Procedia Eng. 114: 685–690. https://doi.org/10.1016/j.proeng.2015.08.010.

Tay, Y.W., Panda, B., Paul, S.C., Tan, M.J., Qian, S.Z., Leong, K.F., et al. 2016. Processing and properties of construction materials for 3D printing. Mater. Sci. Forum. 861: 177–181. https://doi.org/10.4028/www.scientific.net/MSF.861.177.

Teixeira, J., Schaefer, C.O., Maia, L., Rangel, B., Neto, R. and Alves, J.L. 2022. Influence of supplementary cementitious materials on fresh properties of 3D printable materials. Sustainability. 14(7): 3970. https://doi.org/10.3390/su14073970.

Teixeira, J., Schaefer, C., Rangel, B., Alves, J.L., Maia, L., Nunes, S., et al. 2021. Development of 3D printing sustainable mortars based on a bibliometric analysis. Proc. Inst. Mech. Eng., Part L: J. Mater.: Des. Appl. 235(6): 1419–1429. https://doi.org/10.1177/1464420721995210.

Ting, G.H.A., Tay, Y.W.D., Qian, Y. and Tan, M.J. 2019. Utilization of recycled glass for 3D concrete printing: Rheological and mechanical properties. J. Mater. Cycles Waste Manage. 21(4): 994–1003. https://doi.org/10.1007/s10163-019-00857-x.

Turcry, P. and Loukili, A. 2006. Evaluation of plastic shrinkage cracking of self-consolidating concrete. ACI Mater. J. 103(4): 272–279.

United Nations. 2022. The 17 Goals - Sustainable development goals. https://sdgs.un.org/goals.

Chapter **5**

# Ceramic AM and Beyond: The Potential of Hybrid Construction Systems

**João Carvalho*** (https://orcid.org/0000-0002-5650-1221)
**Bruno Figueiredo** (https://orcid.org/0000-0001-8439-7065)
**and Paulo J.S. Cruz** (https://orcid.org/0000-0003-3170-4505)
Lab2PT – School of Architecture, University of Minho

## INTRODUCTION

In the words of Guilherme Wisnik (2016), contemporary architecture is passing through a transformative shift from mechanical form to performative behaviour. This evolution echoes Reyner Banham's notion of architecture serving as a medium for creating "well-tempered environments" (Banham 1969), transcending mere form and structure. The present manifestation of digital architecture embodies the fusion of poetic expression and intuitive knowledge, blending culture, tradition, industry, and progress (Colletti 2010). This harmonious amalgamation seeks to foster a more adaptive and efficient built environment.

---

*For Correspondence: joao.carvalho@eaad.uminho.pt

In the mid-1990s, a notable mention of the potential of Additive Manufacturing (AM) in the field of Architecture emerged as 'Incremental Forming' (Mitchell and McCullough 1997). Commonly known as 3D printing, Additive Manufacturing (AM) consists of producing objects by the successive addition of material layer by layer, where necessary, offering a sustainable and highly cost-effective approach. This is in stark contrast to traditional methods and even contemporary subtractive methods, which typically result in significant material waste during the production processes.

The emergence of AM has sparked a wave of exploration in architectural design, construction materials, and the building industry. In addition, the integration of computational design and simulation tools has unlocked its boundless potential, ushering in an era of exponential growth. By enabling unparalleled geometrical freedom tailored to unique design criteria, optimizing designs, and tackling complex functionalities (Sarakinioti et al. 2020), AM has unveiled new horizons.

Although the concept of AM is rapidly expanding, and the commercialization of 3D-printed houses is already a reality (Casteñeda et al. 2015), there remain two divergent viewpoints on how its effective implementation could unfold. One approach involves applying AM directly on-site, while the other advocates for prefabricating discretized components at the factory and later assembling them on the construction site. The report titled 'Shaping the Future of Construction' (World Economic Forum 2016) clarifies this matter by exploring the potential impact of new technologies on the construction industry. According to this study, the concept of 'contour crafting of buildings' is deemed improbable and to have a relatively low impact. On the other hand, '3D printing of components' and the use of 'advanced building materials' are viewed as moderately likely to occur and to have a moderate impact. Surprisingly, the most promising prospect lies in 'prefabricated building components,' with an exceptionally high likelihood and impact.

This conclusion may explain why numerous research groups and corporations, eager to test AM technologies in construction, tend to adopt an alternative approach centred around discrete elements. As Mario Carpo (2019) points out, 'instead of printing bigger and bigger monoliths, it may conceivably be easier to start with any number of parts, as many and as small as needed', leaving the correct positioning and assembly of the parts to another machine. In this manner, a semblance of traditional building systems is achieved through three key factors: (i) prefabrication of the building's various constituent components into smaller dimensions; (ii) transportation of these components to the work site; and (iii) assembly in their designated positions, resulting in the creation of larger structures.

Ceramics are one of the oldest building materials, dating back to the origins of civilization – clay bricks or adobe were invented between 10,000 and 8,000 BC (Campbell and Pryce 2016). Characteristics such as hardness, density, durability, and the possibility of having a large number of shapes and finishes made ceramics a widely disseminated material (Cruz et al. 2020a). This popularity increased further when it became an affordable material, a result of technological advances that emerged in Europe (Campbell and Pryce 2016). As one of the first materials created by humans, ceramics have a history intricately linked with the history of architecture and construction.

Despite this rich history, the influence of ceramic materials has been diminishing, and their use as a primary construction material has been decreasing over the last few decades, relegating them to a secondary position. However, the possibility of additively producing ceramic components brings new opportunities for the architecture and building industry, opening new possibilities in balancing the use of a low-cost, high-performance material with the execution of complex geometries and multifunctional products. These complexities would be almost impossible to obtain using any other traditional production process (Cruz et al. 2017).

AM can hold a key advantage in revitalizing and rediscovering the potential of ceramic materials in construction and architecture. Leveraging digital design capabilities and the use of high-performance materials through AM opens new possibilities, allowing us to reintroduce traditional ceramic functions that have waned in recent times.

This chapter's focus aligns with these principles, delving into the exploration of digital design and additive manufacturing in ceramics to create a hybrid constructive system. The pursued objectives encompass morphological, technological, and functional explorations of ceramic components. The discretization of structural elements into medium-sized components is undertaken, optimising the use of diverse materials, and also incorporating the principles of Design for Assembly and Disassembly (DFAD). Additionally, our aim is to enhance the functionalities of ceramic components, broadening their scope and applications.

## METHODOLOGY

This work adopts a methodology built upon three essential and interconnected aspects to accomplish the envisioned goals: (i) investigating the properties of ceramic material and potential complementary materials to supplement the primary material. (ii) implementing topological optimization techniques for the structural components, while carefully considering the distribution of material types. (iii) adapting the geometries and dimensions of the components to align with the characteristics of the production equipment.


## MATERIALS

To produce the constructive system ceramic components, we utilized a fine stoneware paste. This material, which is without chamotte and contains 35% of water, proved to be the most suitable and compatible with the type of extruder in our production equipment. Previous studies have demonstrated that this specific material, when fired at 1260°C, can reach up to 175 MPa of compressive strength (Cruz et al. 2020b).

The decision to employ a single material consistently across all prototypes, regardless of their purpose and design, is well-justified. This choice not only stems from the exceptional performance of the material but also streamlines the development process (design), production (printing), and compatibility (assembly) of these components. By utilizing a meticulously stabilized and studied material, we can devote our efforts to refining the design and development of systems rather than struggling to create multiple materials to serve the same purpose. It is essential to acknowledge that the manufacturing method outlined below, and the computational models used for generating the geometries have been optimised specifically for this material. Any alterations to the composition or type of ceramic material would necessitate adapting the method and geometry accordingly.

Considering the production method and its inherent characteristics, especially regarding the size of the components, it becomes necessary to discretize each constructive element into smaller pieces that conform to the equipment's production volume standards. In this context, the concept of connection and union gains crucial significance, becoming an integral part of the project. Whether visibly apparent or not, it can make all the difference between a fully functional system that successfully addresses the intended problem and an impractical and non-functional system that lacks credibility as an alternative to traditional construction methods.

To achieve the desired properties, the bonding material must provide the necessary ductility that ceramics lack while maintaining reasonably high mechanical performance, especially in compression efforts, to maximize the ceramic's potential. Using the previously mentioned stoneware as a base, we considered complementary materials that could serve as connecting elements between various ceramic pieces forming a constructive element. By analysing ceramic specimens produced by Paste

Extrusion Modelling (PEM), simulating real components, we conducted compression tests to classify each material's ability to complement the primary material. Materials tested included wood (oak), rubber (SBR), mortar (sika glue), acrylic glue and concrete (C-30 mixture).

For this purpose, cylindrical ceramic specimens with approximately 70 mm in diameter and 40 mm in height were developed, consisting of three 3 mm thick walls, where the inner wall rises above the exterior ones in height, serving as a connecting element. Later, these pieces were associated with materials that could serve as connection/separation elements.

**Figure 1**  Ceramic specimens for compression tests. From left to right: Single ceramic component; two stacked ceramic components with one layer of separation material; three stacked ceramic components with two layers of separation material.

Based on the aggregation of these components and materials (Figure 1), a series of test specimens were developed, comprising: (i) simple ceramic element; (ii) two stacked ceramic elements; (iii) three stacked ceramic elements; (iv) two ceramic elements stacked with a layer of separation material; (v) three ceramic elements stacked with two layers of the separation material; (vi) two ceramic elements filled inside with concrete; (vii) three ceramic elements filled inside with concrete; (viii) cylindrical test pieces of solid concrete; (ix) cylindrical hollowed concrete test specimens. Three test specimens were developed for each series, making a total of 51 test specimens (Figure 2). The main objective of these tests was to determine which material is best suited to work together with ceramics having as a base point the results obtained by the concrete specimens.

**Figure 2**  Complementary materials specimens and concrete reference specimens for compression tests.

**Table 1**  Complementary materials compression tests

| Test probe | Connection Heigh [mm] | Connection Section Area [mm²] | Average Load [KN] | Standard Deviation [MPa] | Standard Deviation [%] | Average Load [MPa] |
|---|---|---|---|---|---|---|
| Ceramic [1] | 0 | 1870 | 313,1 | 12,9 | 7,7 | 167,4 |
| Ceramic [2] | 0 | 1870 | 120,1 | 15,8 | 24,6 | 64,2 |
| Ceramic [3] | 0 | 1870 | 57,5 | 8,5 | 27,7 | 30,7 |
| Ceramic [2] & Mortar [1] | 3 | 1870 | 78,2 | 8,8 | 21,1 | 41,8 |
| Ceramic [3] & Mortar [2] | 6 | 1870 | 78,2 | 8,3 | 19,9 | 41,8 |
| Ceramic [2] & Acrylic adhesive [1] | 1 | 1870 | 67,7 | 8,1 | 22,4 | 36,2 |
| Ceramic [3] & Acrylic adhesive [2] | 2 | 1870 | 71,5 | 5,1 | 13,4 | 38,2 |
| Ceramic [2] & Rubber [1] | 2,75 | 1797 | 80,3 | 8,4 | 18,8 | 44,7 |
| Ceramic [3] & Rubber [2] | 5,5 | 1797 | 51,0 | 6,4 | 22,5 | 28,4 |
| Ceramic [2] & Wood [1] | 6,08 | 1690 | 165,3 | 5,9 | 6,0 | 97,8 |
| Ceramic [3] & Wood [2] | 12,16 | 1690 | 128,0 | 28,7 | 37,7 | 75,7 |
| Ceramic [2] & Concrete [80] | 1 (acrylic adhesive) | 1870 | 189,0 | 10,8 | 10,7 | 101,1 |
| Ceramic [3] & Concrete [120] | 1 (acrylic adhesive) | 1870 | 124,3 | 12,0 | 18,0 | 66,5 |
| Concrete massive [80] | No connection | 5026 | 129,1 | 6,3 | 24,5 | 25,7 |
| Concrete massive [120] | No connection | 5026 | 107,3 | 3,7 | 17,4 | 21,3 |
| Concrete hollowed [80] | No connection | 2211 | 47,5 | 3,3 | 15,3 | 21,5 |
| Concrete hollowed [120] | No connection | 2211 | 40,3 | 6,4 | 35,2 | 18,2 |

The results obtained from compression tests clearly demonstrate the potential of using this material in structures with a bearing capacity (Table 1). A comparison with concrete, the most used material for supporting structures, reveals a significant difference in capacity between ceramics and concrete. Surprisingly, even without introducing elements to separate ceramic pieces, such as in cases C2 (approx. 80 mm height) and C3 (approx. 120 mm height) with two and three stacked components, respectively, the resistance values (in MPa) were notably higher than those of equivalent concrete specimens CM/CH80 and CM/CH120.

Furthermore, the results from test pieces using rubber (C2R1/C3R2) and acrylic glue (C2A1/C3A2) as separating elements between ceramic pieces indicate instability, leading to pronounced deformations and misshapen settlements, ultimately resulting in total rupture of the ceramic pieces. In contrast, materials like mortar (C2M1/C3M2) and wood (C2W1/C3W2) used in the study show promising results by mitigating some of the weaknesses of the ceramic material and exhibiting higher resistance values than those presented by concrete specimens.

## TOPOLOGICAL OPTIMIZATION

Although AM allows the creation of highly efficient components, where the material deposition only happens where it is strictly necessary, the use of this tool, by itself, does not guarantee that the final component will have these characteristics. As in any other type of manufacturing process, whether additive, subtractive or any other, there must be a direct relationship between the production tool and the design, under penalty of not properly exploiting the advantages of using the various manufacturing processes.

In this sense and in the search for a constructive system that respects the precepts set out for the use of material and structural concerns, topological optimization emerges as a medium to achieve the right balance between form, structure, and material. Rooted in mathematical algorithms, it optimizes material distribution within a given architectural structure to achieve maximum performance while minimizing material usage. By leveraging advanced computational tools like Rhinoceros® and Grasshopper (tOpos plugin), we were able to explore intricate material distributions, uncovering design solutions that enhance both performance and sustainability. When combined with AM, topological optimization elevates the potential of this manufacturing process, ensuring that material is placed precisely where needed, and results in highly efficient and structurally sound components. This approach not only pushes the boundaries of architectural design but also fosters environmentally conscious practices by reducing material waste and promoting resilient structures.

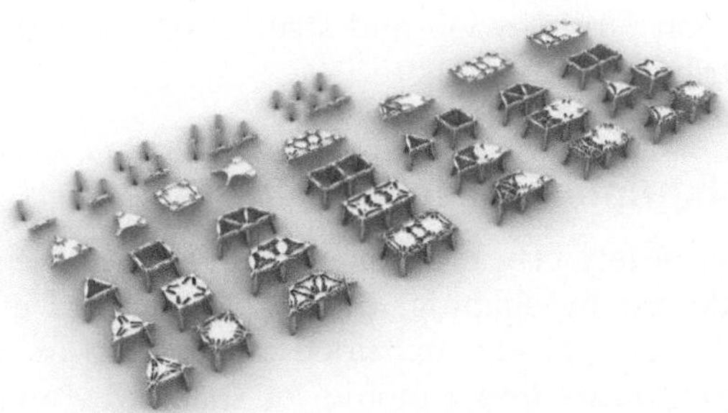

**Figure 3**   Part of topologic optimisations carried out during the investigation, both individual elements and composed systems.

For this investigation, as starting point to define a consolidated and well-informed design, we conducted a topological analysis of a set of standard structures both as a system and individually (Figure 3). Columns, beams, and slabs were examined in various scenarios to fully comprehend the underlying logic of each structural component (Figure 4). Fixed dimensions were established for length, height, and width for each element, maintaining consistency across all tests to facilitate data comparison.

**Figure 4**   Example of the process of topological optimisation applied to a triangular framed structure composed by three columns and three beams. From left to right: Initial boundary surfaces to optimise; Voxel topological optimisation; Mesh topological optimisation.

Upon completing the tests, a thorough analysis of the optimisation results was conducted. The areas requiring material retention, the areas allowing material removal, and the structural scheme with the types of forces involved in each part of the constructive system became evident. Subsequently, we interpreted these results to create a digital model of the overall structure, respecting the distribution of masses and loads within the system.

Design-wise, special attention was given to moments of compression, where ceramic material would predominantly be used. For other mechanical stresses to which the system is subjected, particularly those related to traction forces, wood and steel were selected as materials to efficiently distribute these loads.

## Production Equipment's and Processes

The production of any clay component begins with the creation of the clay body, followed by shaping, drying, firing, and eventually post-processing. Clay is typically extruded with a die in the final cross-sectional shape, poured into a mould, or shaped by an artisan. In any of these production methods, it is necessary to produce multiple parts to offset the initial investment in dies or moulds. One the other hand, when shaping is done manually, the production time for each piece is significant, leading to an increase in the final price of each component. The ability to utilize additive manufacturing for ceramic components presents new opportunities for the building industry to explore and incorporate elements with precise design requirements, reshaping and expanding the boundaries of what can be achieved with masonry construction and ceramic materials.

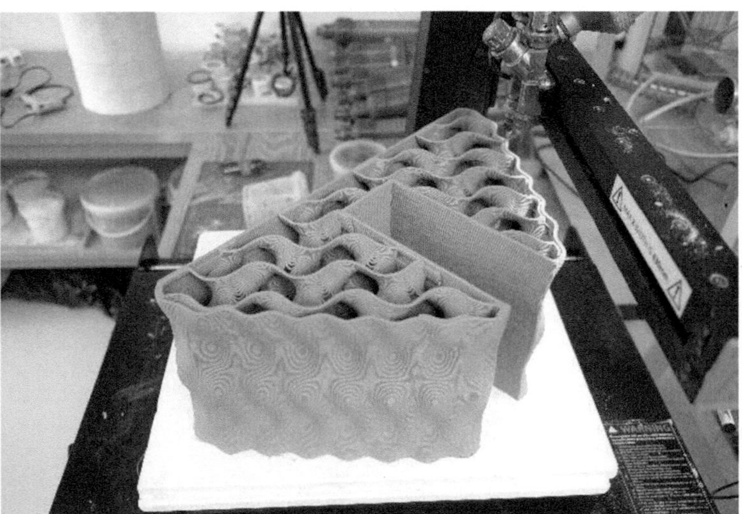

**Figure 5**     Additive manufacturing of a medium size ceramic component with the available production equipment.

The system we employ to produce ceramic components involves a cartesian three-axis printer (Figure 5), where a print head with a motor-controlled rotating spindle is coupled. A cartridge, capable of having various dimensions, is connected to this print head to hold the ceramic material. A tube of compressed air is linked to the cartridge, facilitating the movement of the ceramic material to the rotating spindle. Manual control allows adjustment of speed, extrusion flow, and air pressure during the printing process, ensuring adaptation to the specific characteristics of the clay being used.

However, the primary method of equipment control is through the utilization of a numerical control code (G-Code), which encompasses all the instructions and steps essential to produce an object.

This code condenses pertinent information pertaining to coordinates, travel speeds, printing speeds, paths, and material flow at each moment during the printing process. There are specialized software designed to slice three-dimensional models and generate control codes that work seamlessly with 3D printing standard equipment. It also automatically proposes the integration of temporary support and internal structures when needed, which help reduce material consumption and ease the creation of complex geometries. However, despite their advanced capabilities, these generic software solutions do not permit complete customisation of the numerical control code, a critical aspect necessary to achieve the desired outcome.

Considering this, a computational model was developed in Grasshopper to address the creation of numerical control codes. This model enables the slicing of the initial digital three-dimensional model and facilitates necessary customisations and adaptations, providing the flexibility needed for precise control over the printing process.

In addition to customising the control code and all information within, it was also necessary to adjust the dimensions of each component to ensure they not only fit within the working dimensions of the production equipment – printer and oven – but are also aligned with the available printing times, respecting the labour hours of our laboratory spaces. The production equipment, being relatively small and designed primarily for small and medium-sized components, operates at reduced printing speeds and extruded material volumes compared to the ideal conditions, resulting in smaller components with thinner walls than anticipated.

Consequently, we regard this system as highly suitable for small-scale production or prototyping, which aligns well with our current objectives. However, for industrial-scale production of such components, we recognize the necessity for considerably more robust equipment, such as robotic arms, capable of incorporating extrusion systems that match the dimensions of the components to be produced.

In addition to the above, it is imperative to address the post-printing stage, which holds significant importance. Considering the characteristics of the ceramic material used, known for substantial shrinkage, and the mechanical limitations of the printer necessitating material reduction to avoid lengthy printing times, the resulting component often possesses considerable dimensions and very thin walls. Consequently, during the drying phase, the component is susceptible to dehydration effects. Therefore, it is vital to ensure a uniform and gradual dehydration process to prevent any breakage or distortion of the component (Figueiredo et al. 2019).

Following printing, a post-production phase is conducted before firing, wherein small imperfections arising from the printing process are rectified. Additionally, contact surfaces between pieces are adjusted, aiming to minimize surface irregularities as much as possible when assembling multiple components (Figure 6).

Finally, each of the components is meticulously arranged (Figure 7) and placed inside the kiln, ensuring the most favourable disposition for firing while guaranteeing the integrity of the pieces. It is crucial for the kiln to be fully loaded to optimize energy efficiency during the firing process and to promote smoother heating and cooling cycles, facilitated by the substantial thermal inertia present.

**Figure 6**  Example of post-production of ceramic components: opening of holes for concreting the component.

Firing is a fundamental stage in obtaining components with effective utility. This phase of the process brings about the most significant transformations in the material, affecting both its geometry and molecular structure. Two particularly critical phases occur during the process. At around 573°C, the ceramic paste's quartz undergoes an inversion, transitioning to the high-temperature phase with rapid volume expansion (approximately 2%). Subsequently, between 850°C and

900°C, the ceramic material experiences its most substantial contraction phase, progressing through several stages: formation of clay spinel (950°C); crystallization of clay (980°C); vitrification (1100°C; and fusion of feldspar and clay dissolution (1200°C) (Canotilho 2003).

**Figure 7**　Example of the arrangement of ceramic pieces inside the electric oven (before firing).

It is important to acknowledge that different materials exhibit distinct behaviours, leading to varying final shrinkage values. In the case of the material employed for producing ceramic components in this constructive system, the fixed contraction value after firing stands at 25%, with a firing temperature of 1260°C.

The kiln used for firing the ceramic pieces facilitates the digital creation and adjustment of firing curves, which can be saved in dedicated programs. For the ceramic components produced, changes were made to the original firing curve due to the relationship between their dimensions and volume. Specifically, heating times were increased to mitigate the effects of expansion and contraction phases mentioned earlier. After the firing phase, the components undergo thorough checks for geometric and structural integrity before being approved for assembly.

## CERAMIC HYBRID CONSTRUCTIVE SYSTEM

This investigation was directed towards the development of a constructive system employing ceramic components fabricated through AM (Figure 8), culminated in the prototyping of an extensive range of structural elements. The objective was to explore multiple potential solutions and identify challenges that need to be addressed and mitigated.

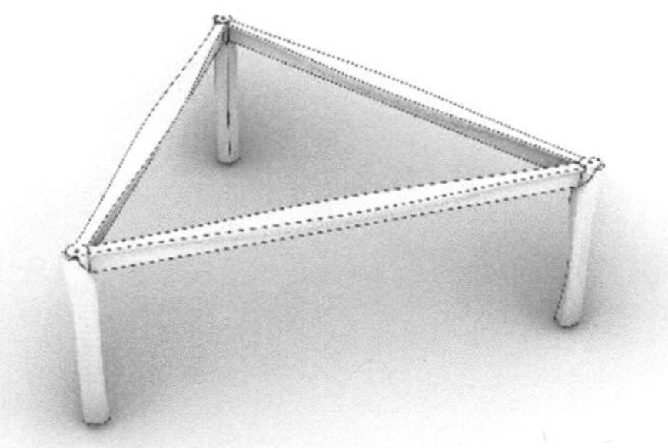

**Figure 8**    Hybrid constructive system.

# Column

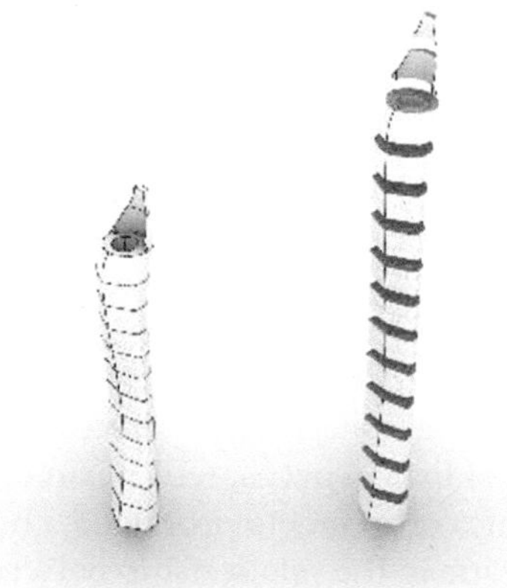

**Figure 9**    Hybrid column.

After completing the tests that demonstrated wood as the most suitable material to complement ceramics, our focus shifted to defining a set of structural elements that could form the basis for studying the main system. This phase started with the design of columns, considering not only their obvious structural role but also additional performative functions (Figure 9).

The selected functions encompassed structural reinforcement, thermal insulation, ventilation, passage of infrastructures, and cladding (Figure 10). At this early stage of the investigation, the functions were chosen to facilitate a seamless transition between constructive systems. The aim was to augment existing systems rather than replacing them entirely. Thus, the ceramic components were designed to enhance and complement existing structures. Consequently, most of the developed components were suitable for use as lost formwork for constructing reinforced concrete columns.

**Figure 10**    Ceramic AM column with thermal insulation.

Moreover, beyond the mentioned functions, the incorporation of these components also significantly elevated the fire resistance levels of the structures they were involved with. This advantageous feature further justifies the application of the ceramic elements, underscoring their positive impact on structural performance and safety.

The final version of the column diverges from the logic of complementarity between existing systems and firmly commits to creating a viable alternative system. Its inception involves the use of a series of ceramic staves (Figure 11), each featuring an internal honeycomb structure with two layers, accompanied by MDF separators to prevent the ceramic components from touching each other. The external shape of the column as well its mass distribution derives from the results obtained from various topological optimization models (Figure 12).

Primarily designed to withstand compression forces, wherein ceramics excel, the column had to accommodate the dynamic nature of structures and the multitude of loads and forces that can act upon it. As a solution, a tensioning element in the form of a steel cable was added to the assembly, compressing it, and making it a monolithic element. This metallic element will also serve to connect with other elements of the system at a later stage.

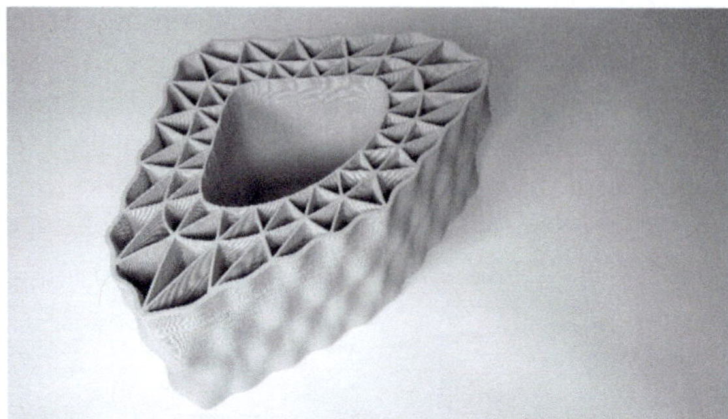

**Figure 11**   A stave of the final hybrid column with its double layered alveolar wall, providing high performance structural capacity. The steel cable that consolidates the column passes through the interior void, making it also possible to pass infrastructures.

**Figure 12**   Surface detail of the hybrid column.

## Beam

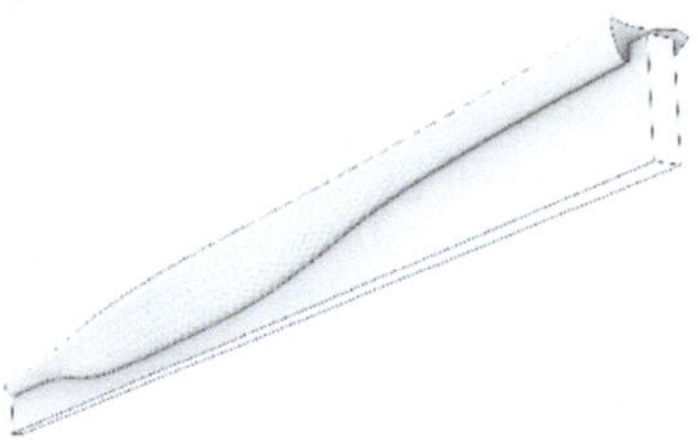

**Figure 13**   Hybrid beam.

With the principles established and upon reviewing the outcomes of the column prototypes' printing and assembly, we move to the next structural element: the beams. In this specific case, the initial approach deviated significantly from the previous prototypes. As a horizontal element, in contrast to the previous vertical one, the construction method would necessarily need to be different and at the same time maintain the use of ceramics as primary material.

Another distinction in the horizontal element lies in the distribution and type of force acting upon the structural element. This variation, evident by the presence of both traction and compression forces in the same element but in different locations, necessitated the incorporation of a new material – steel. As a result, the ceramics are used to withstand compression forces, while wood serves to transfer loads between the ceramic pieces, and the steel effectively resists tensile forces – each material optimally utilized according to its specific strength.

To materialize this concept easily and efficiently, we developed a 2.7-meter-long beam using hollow bricks, MDF boards, and threaded galvanized steel rods (Figure 14). The system's design is straightforward yet effective. The ceramic bricks are aligned and spaced using MDF boards, while the threaded rods serve as connectors to ensure the assembly's coherence and resistance against tensile stresses. Upon completing the assembly, structural testing was carried out by applying weights to the prototype. The results surpassed our expectations, as the prototype displayed no signs of deformation even under the load of three individuals. This successful validation establishes a functional principle, laying the foundation for the subsequent construction of a larger beam.

**Figure 14**   Test of concept using hollowed bricks and steel rods for compression.

For the creation of a functional beam system, we could not confine ourselves to merely applying the concept articulated in the previous prototype. Utilizing the force distribution scheme derived from topological optimization, a novel model was developed, incorporating various materials to enhance performance and address challenges identified in the initial design (Figure 13).

**Figure 15**   Hybrid column before assembly. Wood core and ceramic components.

The primary concern with the preceding system stemmed from the method of connecting the ceramic components, solely relying on friction between the hollow bricks and MDF boards. To address this, a transverse union element was developed for the new model, offering enhanced cohesion and robustness to the assembly. Termed the "core of the beam," this element consists entirely of wood since it comes into direct contact with the ceramic material. Thirty ceramic pieces are then mechanically connected to the core's top (Figure 15).

Ceramic components were specially designed to withstand compressive forces along the longitudinal direction of the beam. Despite the ceramic material's inherent resistance to compressive stresses, an additional layer of strength is achieved by incorporating concrete within the ceramic components. In terms of placement for the concrete, two possibilities exist: it can either serve as a reinforcing element, integrated individually into each ceramic component, or it can serve as an aggregating and unifying element. In the latter scenario, it establishes a direct connection between ceramic pieces and the wooden core through metallic connectors (screws).

**Figure 16**   Hybrid beam.

The primary distinction between these two versions lies in the ease of assembly reversibility and the straightforward replacement and repair of individual beam components, a feature exclusive to the first version.

To optimize ease of assembly and ensure compatibility with available production equipment, the beam prototype employs the second version, making it infeasible to disassemble the set once assembled.

No tests were conducted to determine the maximum resistance of this element, and no prototypes were created using alternative construction systems for direct comparison. However, it is important to emphasize that there was a significant weight reduction of over 50% when compared to a reinforced concrete beam with similar dimensions (Figure 16).

## Connections between Columns and Beams

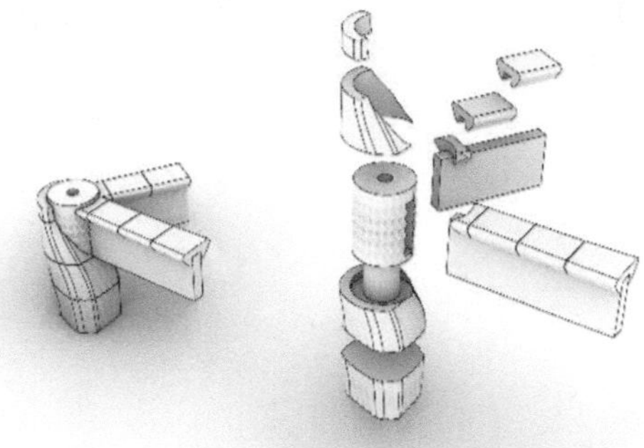

**Figure 17**    Connection between columns and beams.

The connections between vertical and horizontal elements are arguably the most critical and sensitive aspects of a framed structure (Figure 17). Apart from facilitating the transfer of loads to the interconnected elements, these connections play a crucial role in ensuring the overall stability of the structure. By effectively restraining rotational forces in multiple directions, the connections contribute to the structural integrity of the entire assembly.

This preponderance of stabilizing the constructive system becomes even more evident in structures composed of several independent elements, where the structure does not present a monolithic behaviour and where the possibility of disassembling the system is foreseen, as is the case of the constructive system studied and developed within the scope of this investigation. In the context of the developed constructive system, which incorporates discretized ceramic components alongside the complementary materials, the connection between vertical and horizontal elements plays a crucial role in solidifying the entirety of the system. It is also in this moment of meeting of directions that the structural consolidation is carried out element by element, since all these composite elements comprise the post-tension of the components, which has the fixings at the ends of the elements, increasing the complexity of the connection.

To formalise a functional connection between vertical elements, a range of connection concepts was developed, catering to different structural conditions, including variations in structural blocking and connection types. To achieve this, various connection types commonly employed in traditional construction systems were considered. These connections were categorized into four structural schemes, each encompassing four types of structural blocking. After evaluating the various options, we opted to proceed with the prototyping of one of these configurations, while incorporating adaptations to the column and beam designs.

**Figure 18**   Connection between one column and two beams.

The component responsible for connecting the columns and beams is entirely constructed from wood, ensuring enhanced flexibility and ductility in the union. Both geometrically and functionally, this component is relatively straightforward and easily achievable (Figure 18). It comprises the overlapping of two cylinders with varying diameters, featuring a longitudinal hole and a set of cavities specifically designed to accommodate the body of the beams.

The connection to the columns is established through its last two components, where each connection cylinder is positioned. The central hole in these cylinders accommodates a steel cable, which serves to reinforce the column and counteract lateral movements arising from the compression of its components.

As for the beams, their connection is achieved through cavities specifically designed for this purpose. Additionally, it is possible to connect these elements via a steel cable passing through the interior of the beam. This cable will connect to each of the wooden connectors located at the beam's ends, further enhancing the structural integrity of the overall system. In this manner, the beam and column are combined into a single element.

## Slabs

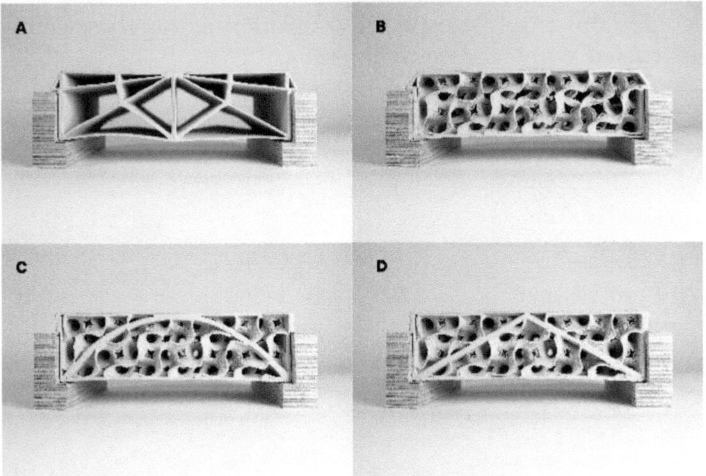


**Figure 19**   Types of specimens for load-bearing tests.

Concluding the set of building elements in a typical building system, we encounter the slabs, which function as the horizontal, planar elements encompassing the floors and roofs of structures. Like the previously mentioned components, conventional building systems involve scenarios in which ceramic materials are utilized, either entirely or partially, in the construction of these essential structural elements.

Prior to formalizing any system for constructing slabs, and in line with the methodology employed in the previously presented architectural elements, a series of components with varying geometries and different operational approaches were considered for preliminary analysis. The objective was to evaluate the potential of each typology and establish a path forward during the design phase of the final system.

Consequently, four types of components were conceived (Figure 19), featuring variations in mass organization and structural schemes. The design of these components draws inspiration from traditional ceramic vaults and adheres to their operating scheme, necessitating lateral supports (beams) for stability and reinforcement.

The support element employed to assess the maximum capacity of each geometry consists entirely of wood and follows the section design of traditional prestressed beams. All component types comprise two lateral supports, an upper, and a lower one.

The results of the mechanical tests exhibit significant variations in the average capacity among the different types of vaults, with notable benefits observed in components where structural reinforcement and formal integrity logics were introduced, particularly in types C and D. Type A components achieved an average capacity of 10 kN, while type B components reached an average capacity of 7 kN. These values are considered highly positive, especially considering that these components are constructed with 1.5 mm thick walls.

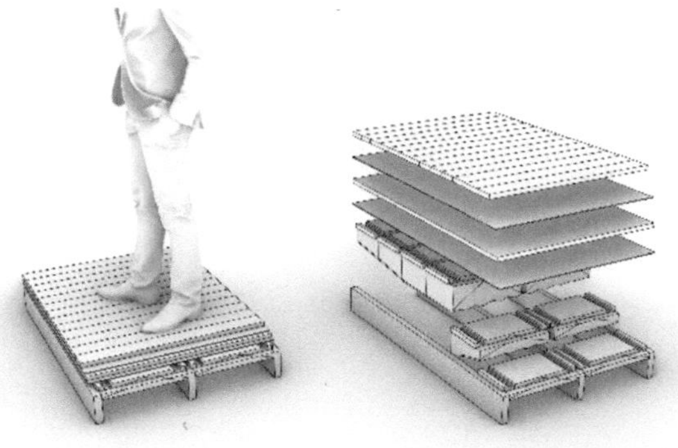

**Figure 20**   Hybrid slab.

As for types C and D, contrary to initial expectations, they exhibited mechanical resistances significantly higher than those of the previous types. Even when subjected to the maximum admissible capacity of 45 kN (90% of the equipment's maximum capacity), it was not possible to reach the limit of these components in the first phase. All specimens of types C and D were tested with the wooden base, applying a force up to the maximum admissible value of 45 kN.

Considering the fact that the ultimate objective of these tests is to determine the response of the analysed components to compressive forces, it was deemed valuable for the investigation to subject each specimen to failure by pushing them to their limits. In this regard, all specimens related to types C and D were retested using the same equipment. During this second series of tests, certain parameters of the mechanical tests were altered.

Previously, wooden elements were used to simulate the beams supporting the vaults, serving to absorb and standardize surface tensions. To create more unfavourable conditions for the ceramic material and consequently reduce its mechanical strength, we replaced the wooden elements with steel ones. The final test results reveal that type D demonstrated the highest stability and resistance among all the components under examination.

Once the testing phase concluded, we proceeded to produce a section of the slab, which would later be assembled along with the remaining prototypes. The design of the slab, like other constructive elements comprising the system, adheres to the principles of mass distribution established in the topological optimization models. However, there is still room for rationalization and simplification of the geometries to ensure compatibility with the production equipment.

The slab system is based on the concept applied to beams. It employs longitudinal wooden elements that rest directly on the beams, forming the foundation for placing ceramic vaults, imitating the operational arrangement of conventional lightweight slabs. If necessary, the wooden elements may incorporate post-tensioning elements to enhance the system's mechanical stress response without having to significantly enlarge the dimensions of the slab components (Figure 20).

Once the joists and vaults are assembled, a cork plate (approximately 5 mm thick) is placed over the latter. Subsequently, wooden boards are attached using screws to consolidate the structure. The desired flooring is then installed on top of these wooden boards.

## RESULTS AND DISCUSSION

**Figure 21**    Hybrid construction system.

The developed system (Figure 21), along with its underlying principles and the set of practical outcomes achieved during the prototyping of diverse models, reinstates the belief in the viability and feasibility of manufacturing medium-sized ceramic architectural components with PEM. These systems and components can subsequently be integrated into real context, reshaping the built environment to be more cohesive and environmentally conscious.

The system introduced in this study represents a significant advancement in the pursuit of more environmentally sustainable construction methodologies tailored to specific contexts. Each individual element can be tailored for distinct performance attributes, forms, or functions, ensuring differentiation while being seamlessly integrated within a cohesive framework.

By allocating materials according to the structural arrangement derived from topological optimization, a hybrid system is established, ensuring optimal material performance under distinct forces. Additionally, the application of design principles for assembly and disassembly facilitates a fully reversible and remarkably repairable system. Damaged components can be seamlessly replaced without substantial limitations. Consequently, the integration of these materials and processes, spanning from design conception to production, culminates in a constructive system that judiciously governs the choice and volume of materials employed, thereby making a substantial stride toward cultivating a more sustainable built environment.

The results obtained from the compression tests unmistakably demonstrate the material's significant potential for application in load-bearing structures. Comparing the attained values with those of concrete, the widely employed material for constructing support structures, underscores the substantial capacity contrast between ceramics and concrete, with ceramics having a considerable advantage.

In any case, it is important to acknowledge the inherent challenges associated with this type of production, primarily linked to the material properties and their diverse responses throughout the production process.

The system presented in this study was developed within a laboratory and experimental context. Transitioning from such a system to practical application and effective construction of structures necessitates fundamental adjustments. As previously mentioned, the production equipment employed is specifically designed for small and medium-scale components, as well as limited production quantities.

Furthermore, the ceramic material used poses challenges in controlling its different phases, particularly during the drying and firing stages. To facilitate the system's utilization, alterations would need to be considered. For the scope of this investigation, fine stoneware without chamotte was selected due to its heightened adaptability to the available extrusion system.

When considering larger-scale components, it is recommended to opt for a material with a lower degree of shrinkage and subsequent deformation. This choice helps mitigate potential issues associated with size alterations during the drying and firing processes, ensuring the structural integrity of the final components.

## ACKNOWLEDGEMENTS

This work was financed by the project Lab2PT – Landscapes, Heritage and Territory laboratory, reference UIDB/04509/2020 through FCT – Fundação para a Ciência e a Tecnologia and the FCT Doctoral Grant with the reference SFRH/BD/138062/2018. We are also grateful to Instituto de Design de Guimarães for hosting and supporting the activities of the Advanced Ceramics Laboratory on the use of their facilities and equipment.

## REFERENCES

Banham, R. 1969. The Architecture of the Well-tempered Environment. London: The Architectural Press.

Campbell, J. and Pryce, W. 2016. Ladrillo. Historia Universal. Art Blume, S.L.

Canotilho, M.H.P.C. 2003. Processos de cozedura em cerâmica. Instituto Politécnico de Bragança.

Carpo, M. 2019. The age of computational brutalism. pp. 8–9. *In*: Robotic Building: Architecture in the Age of Automation (1st ed). Munich: Detail Business Information GmbH.

Castañeda, E., Lauret, B., Lirola, J.M. and Ovando, G. 2015. Free-form architectural envelopes: Digital processes opportunities of industrial production at a reasonable price. J. Facade Des. Eng. 3(1): 1–13. doi:10.7480/jfde.2015.1.914.

Colletti, M. 2010. Digitalia – The Other Digital Practice. Architectural Design, 80 (2) - Exuberance: New Virtuosity in Contemporary Architecture, 16–23.

Cruz, P.J.S., Knaack, U., Figueiredo, B. and Witte, D. De. 2017. Ceramic 3D printing – The Future of Brick in Architecture. Proceedings of the IASS Annual Symposium 2017, IASS 2017 – Interfaces: Architecture. Engineering. Science. Hamburg: HafenCity University.

Cruz, P.J.S., Figueiredo, B., Carvalho, J. and Campos, T. 2020a. Additive manufacturing of ceramic components for façade construction. J. Facade Des. Eng. 8(1): 1–20. https://doi.org/10.7480/jfde.2020.1.4725.

Cruz, P.J.S., Camões, A., Figueiredo, B., Ribeiro, M.J. and Renault, J. 2020b. Additive manufacturing effect on the mechanical behaviour of architectural stoneware bricks. Constr. Build. Mater. 238: 117690. https://doi.org/10.1016/j.conbuildmat.2019.117690.

Figueiredo, B., Cruz, P.J.S., Carvalho, J. and Moreira, J. 2019. Challenges of 3D printed architectural ceramic components: Controlling deformations and cracks. Proceedings of the IASS Annual Symposium 2019 – Structural Membranes 2019.

Mitchell, W. and McCullough, M. 1997. Digital Design Media. New York: John Wiley & Sons, Inc.

Sarakinioti, M-V., Konstantinou, T., Turrin, M., Tenpierik, M., Loonen, R., Klijn-Chevalerias, M.L., et al. 2020. Development and prototyping of an integrated 3D-printed façade for thermal regulation in complex geometries. J. Facade Des. Eng. 6(2): 029–040. doi:10.7480/jfde.2018.2.2081.

World Economic Forum. 2016. Shaping the Future of Construction - A Breakthrough in Mindset and Technology, Industry Agenda, Ref. 220416, Prepared in collaboration with The Boston Consulting Group. Switzerland: World Economic Forum.

Wisnik, G. 2016. Inmersión contra imagen: desenfocando el mundo. In Plot nº32, August–September, Buenos Aires: Grupo Vórtice, 2016. p. 10–12.

# Index